Elephant Seals

Pushing the Limits on Land and at Sea

How did the elephant seal survive being driven to the brink of extinction in the nineteenth century? What variables determine the lifetime reproductive success of individual seals? How have elephant seals adapted to tolerate remarkable physiological extremes of nutrition, temperature, asphyxia, and pressure?

Answering these questions and many more, this book is the result of the author's 50-year study of elephant seals. The chapters cover a broad range of topics including diving, feeding, migration, and reproductive behavior, yielding fundamental information on general biological principles, the operation of natural selection, the evolution of social behavior, the formation of vocal dialects, colony development, and population changes over time. The book will be a valuable resource for graduate students and researchers of marine mammal behavior and reproductive life history as well as for amateur naturalists interested in these fascinating animals.

Bernard J. Le Boeuf is Emeritus Professor of Biology at the University of California, Santa Cruz. He is considered one of the pioneers of the field of marine mammal behavior, known particularly for his studies of seal social behavior, diving behavior, diving physiology, and migration. He has published widely on topics in reproductive behavior, ecology, and behavioral biology.

Elephant Seals

Pushing the Limits on Land and at Sea

BERNARD J. LE BOEUF
University of California, Santa Cruz

CAMBRIDGE
UNIVERSITY PRESS

University Printing House, Cambridge CB2 8BS, United Kingdom

One Liberty Plaza, 20th Floor, New York, NY 10006, USA

477 Williamstown Road, Port Melbourne, VIC 3207, Australia

314–321, 3rd Floor, Plot 3, Splendor Forum, Jasola District Centre, New Delhi – 110025, India

103 Penang Road, #05–06/07, Visioncrest Commercial, Singapore 238467

Cambridge University Press is part of the University of Cambridge.

It furthers the University's mission by disseminating knowledge in the pursuit of
education, learning, and research at the highest international levels of excellence.

www.cambridge.org
Information on this title: www.cambridge.org/9781316511541
DOI: 10.1017/9781009052085

First published 2021

A catalogue record for this publication is available from the British Library.

Library of Congress Cataloging-in-Publication Data
Names: Le Boeuf, Bernard J., author.
Title: Elephant seals : pushing the limits on land and at sea / Bernard J. Le Boeuf, University of California,
Santa Cruz.
Description: Cambridge, United Kingdom ; New York, NY : Cambridge University Press, 2021. |
Includes bibliographical references and index.
Identifiers: LCCN 2021017370 (print) | LCCN 2021017371 (ebook) | ISBN 9781316511541 (hardback) |
ISBN 9781009055031 (paperback) | ISBN 9781009052085 (epub)
Subjects: LCSH: Elephant seals. | Elephant seals–Adaptation.
Classification: LCC QL737.P64 L425 2021 (print) | LCC QL737.P64 (ebook) | DDC 599.79/4–dc23
LC record available at https://lccn.loc.gov/2021017370
LC ebook record available at https://lccn.loc.gov/2021017371

ISBN 978-1-316-51154-1 Hardback

Contents

Color plates can be found between pages 118 and 119.

Preface

The elephant seal is associated with superlatives: the largest seal with males weighing over two tons; extremely polygynous, with some males mating with hundreds of females; and the most sexually dimorphic marine mammal, with males being 3–10 times larger than females. Females fast while lactating. Alpha males fast for over a hundred days during the breeding season. They dive deeper and longer than other seals and sea lions, and they spend more time at sea submerged than most whales. No other large mammal has come so close to extinction and made such a rapid and successful recovery. These animals are fascinating because they stretch the boundaries of what can be done both on land and at sea. This book describes the essential elements of the behavior and biology of northern elephant seals that have enabled them to thrive and adapt on land and at sea.

We know a great deal about this animal because it has been studied thoroughly, intensively, and continuously at the same site, Año Nuevo State Park, for over five decades. More than 200 peer-reviewed research papers have been published on the elephant seals from this rookery. Two reasons account for the duration and abundance of the studies. First, these seals are exceptionally amenable to long-term study because individuals can be identified and followed for life; they are robust against disturbance and do not flee in fear; they breed in the open air where they can be viewed; they can be tagged, weighed, measured, blood-sampled, and tissue-sampled; and they will carry attached recording instruments for up to eight months at sea without harm and effects on their subsequent behavior. Second, they inhabit and breed in nature a mere 25 minutes away from the University of California at Santa Cruz, where researchers can sally forth with minimal effort and cost to conduct studies daily throughout the year and continuously over the years.

The perspective here is my own. I describe studies by me and my students and colleagues conducted over five decades on a wide variety of research projects on the behavior, physiology, and biology of these animals in the field. Most of these studies have been published in peer-reviewed scientific journals. Some of these studies are long term, involving the tracking of individuals over their lifetimes. Firsthand experience permits additional insights that enable presentation of personal stories and reminiscences in the text (boxes) that didn't fit in research articles or book chapters, the main source of reporting for academics. Our research on northern elephant seals is compared with findings from several research teams studying the southern elephant seals in the southern hemisphere.

For the sake of making the story more accessible to a wide audience and for ease of reading, I've minimized the use of references, tables, and figures, and I've excluded equations, statistical tests, and mathematical models altogether. These details can be tracked down in the references cited. This book should be of interest to students of animal behavior, marine mammalogists, and naturalists, and it should be especially useful and informative to visitors to marine mammal parks and aquaria and the thousands of tourists that view elephant seals in nature annually at mainland sites in California and Argentina.

For those who want to consult original sources or want more detail than is provided in the text, consult the References. The list includes published peer-reviewed research articles, books, book chapters, and popular pieces, with a special emphasis on the research conducted at Año Nuevo State Park.

The metric system is used throughout this book with length measurement in meters (m) or kilometers (km) and temperature in Centigrade units (°C).

Acknowledgments

The research addressed in this book was conducted by a cast of thousands. This is not a Rabelaisian estimate. Colleagues, collaborators, postdoctoral researchers, graduate and undergraduate students, benefactors, and friends all contributed to the advancement of our knowledge of the elephant seal. I am especially grateful to my colleagues Richard Peterson, Daniel Costa, Leo Ortiz, Carl Hubbs, and Yasuhiko Naito for helping to get things started, providing expertise that I lacked, and providing inspiration and innovation that determined the next questions. I learned much from Robert Trivers, James Estes, Gerald Kooyman, and Terri Williams that influenced our research. I was privileged to work with students and colleagues that heightened the quality of our research: Daniel Crocker, Richard Condit, Joanne Reiter, Marianne Riedman, Patricia Morris, Claudio Campagna, Patrick Robinson, and Susanna Blackwell, all of whom have continued to conduct research on their own. It was a pleasure to work with Phil Thorson, Charles Deutsch, Mike Haley, Walter Clinton, Maria Kretzman, David Aurioles, Naomi Rose, Sara Mesnick, Cathy Cox, Tony Huntley, Stacia Fletcher, Guy Oliver, and Paul Webb. I thank numerous collaborators for their valuable expertise: Rus Hoelzel, Cathy Debier, Sheila Thornton, Pete Klimley, Randall Davis, Russ Andrews, Robert Laws, Wataru Sakamoto, Fritz Trillmich, and AnaValenzuela-Toro. I am grateful to colleagues at other nearby institutions that provided valuable information on sightings of our tagged animals, especially David Ainley, Harriet Huber, Sarah Allen, Peter Pyle, William Sydeman, and Brian Hatfield, among others. Our research could not have been done without long-term and continuous funding from the National Science Foundation and grants from the Minerals Management Service (or Bureau of Land Management), the National Geographic Society, the Office of Naval Research, the University of California Institute for Mexico and the United States, and the University of California at Santa Cruz. We are grateful for philanthropic assistance from George Malloch. I thank the staff at Cambridge University Press, especially Dr. Dominic Lewis, Megan Keirnan, Aleksandra Serocka, Jenny van der Meijden, and Franklin Mathew Jebaraj, for their invaluable help and expertise in the process of publishing this book.

1 Origins, Misnomers, and Bottleneck

The language left some room for confusion.

Anonymous

An animal is virtually nonexistent until it has a name. It is simply an "it" or the group of animals is "them." Our ancestors living near the sea had many names for seals: sea dogs, sea cats, sea wolves, sea bears, sea lions, whale horses, and sea elephants. Names are important, and the name for elephant seals or sea elephants has a peculiar history (Le Boeuf and Laws 1994). As is obvious to anyone that has seen these animals, the outstanding characteristic of a male elephant seal is that it is large and has an exceptionally long nose that falls into its open mouth when it issues a vocal threat to a competitor (Figure 1.1). Someone suggested that the animal should have been named after a politician, but this got nowhere, but who knows, this proposal might surface again.

If naming animals were practical, a suitable name would have alluded to the fact that it was a Big Seal with a Big Nose, or in the Linnaean classification system specifying genus and species, *Megaphoca macrorhinus*. But this was not to be. Linnaeus (1758) had his rules and one of them was precedence. *Macrorhinus* was proposed in 1824 (Geoffroy Saint-Hilaire and Cuvier 1826) but the name had already been taken by a genus of coleopterous insects, and a few years later, the species name was given to the sliteye shark, *Loxodon macrorhinus* (Müller and Henle 1839). Moreover, *Elephas maximus* was already taken by the Asian elephant (Linnaeus 1758).

So, *Mirounga*, from *Miouroung* – an old Australian Aboriginal name for elephant seals – was proposed and has been the genus name since 1827 (Gray 1827). The species name *angustirostris*, or narrow nose, was proposed in 1866 (Gill 1866) to distinguish the northern species from the southern species, *M. leonina*. The proposer, an anatomist, was impressed – as was most of the classifiers in the nineteenth century – by the peculiarly narrowed and pronounced snout of the only northern specimen he had at his disposal, a female skull. What an irony! It is the snout of the male that is distinctive (Figure 1.2). Moreover, the name of the southern species, *leonina*, meaning lion-like, was based entirely on skull fragments and an erroneous description by Lord Anson of a sea-lyon from islands off the coast of Chile (Anson 1748). The southern species bears absolutely no resemblance to a lion or a sea lion. So, we are left with misnomers for both the northern and the southern elephant seal. The Linnaean system of classifying and naming organisms originated in 1735 and is based on similarities in

Figure 1.1 An adult male surrounded by pregnant and nursing females emits a threat vocalization to a competitor from the stereotypical posture. Note that he is squashing a pup that is crying out in distress. (A black and white version of this figure will appear in some formats. For the color version, please refer to the plate section.)

Figure 1.2 Males and females are easily distinguished from facial features or size. (A black and white version of this figure will appear in some formats. For the color version, please refer to the plate section.)

physical traits. The system continues to be used today, with modifications, but can appear arbitrary, frivolous, and misleading at times. Perhaps the geneticists and evolutionists, who are currently displacing anatomists in the taxonomic naming game, will correct these oddities and contradictions.

Figure 1.3 Both species of elephant seals are similar in appearance. Top: headshots of an adult male northern elephant seal on the left and an adult male southern elephant seal on the right. Bottom: adult female and male northern elephant seals on the left and adult female and male southern elephant seals on the right. (A black and white version of this figure will appear in some formats. For the color version, please refer to the plate section.)

Elephant seals are the largest of the 34 species of pinnipeds – the feather- or fin-footed, semiaquatic, carnivorous marine mammals. They are phocids, true seals, as distinct from otariids, fur seals and sea lions, and odobenids, represented by the walrus. There are two species of elephant seals in the genus *Mirounga*: northern and southern (Figure 1.3).

The evolutionary origins of elephant seals and how they expanded into both hemispheres are obscure (Riedman 1990). Phocids are thought to have originated in the North Atlantic 15–20 million years ago and invaded the southern hemisphere about 10 million years ago. Most collections are, however, from the northern hemisphere, which biases interpretations of evolutionary history (Valenzuela-Toro and Pyenson 2019). Recent collections in Chile and New Zealand are forcing a revision of evolutionary history; these findings indicate that rather than diversifying in the North Atlantic (Briggs and Morejohn (1975), monk seals and elephant seals evolved in the southern hemisphere and later reinvaded the northern hemisphere (Valenzuela-Toro et al. 2013; Bossenecker and Churchill 2016; Rule et al. 2020).

Today, northern elephant seals are distributed along the west coast of North America from mid-Baja California, Mexico, to the eastern Aleutian Islands in Alaska. Southern elephant seals inhabit island and mainland sites around the Antarctic continent and in

Patagonia, Argentina. The distance separating the breeding ranges of the two species is 4,000–8,000 km. There is no evidence of intermingling, but vagrant southern elephant seals have appeared in the northern hemisphere as far north as the state of Indiana, the Arabian coast (Sultanate of Oman), the Gulf of California (Sea of Cortez), and the Gulf of Panama, Taboga Island (Johnson 1990; Redwood and Felix 2018; Elorriaga-Verplancken et al. 2020; Valenzuela-Toro et al. 2020). To my knowledge, vagrant northern elephant seals have not been observed south of the equator.

It is not clear how long the northern and southern species have been separated. Separation may have been as recent as the last major glaciation, which ended 5,000–10,000 years ago, or as long ago as the early Pleistocene. The two species are distinct in some ways and similar in others (Le Boeuf and Petrinovich 1974a; Le Boeuf and Laws 1994). The southern species is larger, especially the males, and can bend backwards, assuming an extreme U-shaped posture. Males of both species possess a long proboscis and a rugose chest shield of thick skin, which receives most of the blows and bites during fights. The nose is not involved in threat vocalizations of males, as was once thought. These secondary sexual characteristics signal that the bearer is an adult male; they are more pronounced in the northern than the southern species. The two species live in different environments. The threat vocalizations of the males sound different but have a similar function. These differences imply a separation long enough for the evolution of different structures. Nevertheless, their aquatic and breeding behaviors are remarkably similar (Figure 1.4).

If nomenclature were closer to reality and elephant seals were named *Megaphoca macrorhinus*, the two species might be further differentiated as *boreus* for northerns and *australis* for southerns. Not only would this be the most fitting description, but it has a poetic ring. Perhaps, then, local sports teams along the west coast of the United States would be more likely to take the seals as mascots, instead of the overused

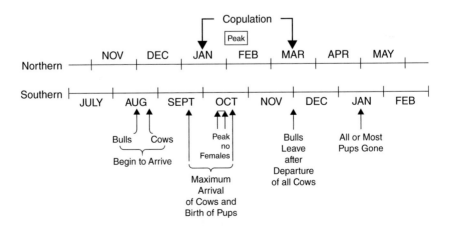

Figure 1.4 The major events in the breeding season of northern and southern elephant seals are similar, with both occurring during the coldest months of the year.

bulldogs, eagles, lions, tigers, wildcats, cougars, panthers, and so on, which have no local connection. Can't you just hear the cheerleaders urging and cheering on the crowd with chants like this:

> Mega-phoca, macro-rhinus
> Rah, rah, rah,
> Smack 'em in the rostrum,
> Sis, boom, bah.

After all, if the University of California at Santa Cruz can adopt the banana slug, *Ariolimax californicus*, as its mascot and get free publicity with John Travolta wearing a tee shirt emblazoned with a bright yellow banana slug in the movie *Pulp Fiction*, it is no great stretch for a school to use Megaphocs as its brand. Go Megaphocs!

Whatever you want to call them, northern elephant seals (*Mirounga angustirostris* or *Megaphoca macrorhinus boreus)* are the focus of this book. Experts agree that they were present in California waters by the late Pleistocene. Archaeological remains confirm that they were here when humans colonized the region over 15,000 years ago. Remains in kitchen middens suggest that elephant seals and other pinnipeds were important for the subsistence of aboriginals in the area at the time (Glasgow 1980).

Certainly, elephant seals were prominent and abundant along the entire coast from southern Alaska to mid-Baja California in the latter part of the eighteenth century. We don't know what their population was when seal hunting began in the 1800s, but judging from the seal harvests that followed, it was likely in the range of 100,000–300,000 (Busch 1985). There are records of elephant seals harvested at Los Coronados, Bahia San Cristobal, Islas San Benito, Isla Cedros, and Isla de Guadalupe in Mexico, at Santa Barbara Island in southern California, and at the Farallones, near San Francisco Bay, in central California. It is likely that they were harvested at most places where they breed today. A whaling captain, Charles Scammon, recalled that elephant seals used to breed from Point Reyes, just north of San Francisco Bay, south into Mexican waters as far as Cedros Island, Cabo San Lázaro, and Guadalupe Island (Scammon 1874). We got reports of the whereabouts and abundance of elephant seals, however, only from the merchants whose business was killing them for profit.

The early 1800s to the 1860s was a disastrous time for elephant seals and most other marine mammals along the west coast of the United States. It was theriocide, the slaughter of wild animals on a grand scale (Beirne 2018). Elephant seals, sea otters, sea lions, and whales were harvested commercially and massacred relentlessly to such an extent that their populations were virtually annihilated. First it was for the pelage of sea otters and fur seals, then it was for the blubber of whales and elephant seals that was rendered to oil. The crews on many vessels were opportunists and took whatever marine mammals were available. One estimate of the number of elephant seals harvested is 250,000 but this is an arbitrary figure from few and poor records that are difficult to interpret. One thing is certain: "elephanting" was a sordid business. As one wag described it: "One need not sentimentalize over sea elephants. Their only use to the world is to provide blubber." This reflects the prevailing philosophy of the day that living things were feckless unless they were useful for humans. Of course, profit justified everything; fur commanded a handsome price in the Orient and oil was a necessary commodity at home because the growing human population needed it. Petroleum would come later. Animal oil, and later tallow, served many needs of the times. Seal oil was a better grade of oil than most whale oils; it burned with a clearer flame at home and in street lighting, and it had wide use in lubrication and other necessities such as soap, paint, oils for tanning, and finishing leather. A big bull could produce as much as 210 gallons of oil. Moreover, the seals were easy to slaughter. They were unafraid of humans and didn't flee into the water like fur seals and sea lions, and they were not as difficult and dangerous to slaughter as gray whales. A bull might sleep while its neighbor was shot and flensed of its blubber on the spot, and the sleeping seal was next. The seals could be herded and killed at leisure. Although large and formidable, they were easy pickings for men with guns or lances.

By the end of the Civil War, there were so few left that it was no longer commercially profitable to hunt them (Stewart et al. 1994). From about 1865 to 1892, only a few were seen in Mexico or no elephant seals were seen at all. The species was considered extinct by the late 1870s; this was before the northern elephant seal was considered a different species from the southern elephant seal. In 1880, a small herd was discovered at Bahía San Cristobal in Baja California. Over the next four years, all 335 of these seals were killed. In 1883, 80 elephant seals were killed at Isla de Guadalupe, a remote volcanic island located in the Pacific Ocean 150 miles west of the Baja California mainland. Another four were killed there in 1884. The species was again considered extinct.

In 1892, however, C. Townsend and A. Anthony, on a collecting expedition for the Smithsonian Institution, saw eight elephant seals at Isla de Guadalupe (Townsend 1912). The collectors were elated and rejoiced at seeing these survivors, and then, with nary the slightest pause, they killed seven for the Smithsonian's collection. Of course, they realized that these animals represented the last of an exceedingly rare species but adding to the collections was their highest priority. Indeed, the rarity of the seals made the collections even more important. This was the "science" of the day.

The late 1890s were clearly the nadir in the long history of the species. Indeed, the species was one step away from the plunge into the abyss of extinction. Remote Isla de Guadalupe was the last refuge. The seals went through a population bottleneck, meaning that a population of thousands of seals was reduced to a few survivors. It was estimated that fewer than 20 animals survived (Bartholomew and Hubbs 1960). This speculation was later confirmed by an investigation of the DNA sequence diversity in two mtDNA regions (Hoelzel et al. 1993; Hoelzel 1999). The genetic studies showed that there had been one or more population bottlenecks, or genetic bottlenecks, and calculated their size and duration. The conclusion was that there was an extreme bottleneck, perhaps lasting 20 years, where only 10–20 seals remained in the population. The overall genetic diversity of the species had been greatly reduced. The survivors lost genes and gene combinations. They were remarkably homogeneous in an allozyme survey (Bonnell and Selander 1974; Hoelzel and Le Boeuf 1990; Hoelzel et al. 2002) as well as in mtDNA analysis. In contrast, the southern elephant seal, which had been commercially harvested as well but not to the same degree as the northern species, did not lack genetic diversity.

There is no doubt that the northern elephant seal population was squeezed through an extreme population bottleneck that changed its genetic structure forever. Think what this means. A population or genetic bottleneck reduces variation in the gene pool; genes or gene combinations that took eons to accumulate are lost. The survivors have less genetic diversity to pass on genes to future generations of offspring through sexual reproduction. Genetic diversity remains low, increasing very slowly over time as random mutations occur, only some of which are favorable, and this is a very slow process. The robustness of the pre-bottleneck population, selected and formed over a long time, was compromised and the survivors may be incapable of adapting to environmental changes, such as a shift in available resources. After all, the best hedge against an uncertain future is genetic variation; that is what sexual reproduction brings to the table. That is not all; genetic drift and inbreeding are associated with population bottlenecks and can lead to further loss of alleles or an increase in deleterious mutations. Numerous studies have shown that there is a positive correlation between genetic diversity and measures of fitness.

Are the elephant seals of today different from those that existed before the population bottleneck? Probably, but this is difficult to confirm, and if they are different, it is not clear how the differences are manifested. Although the elephant seals of today lack genetic diversity, there is no doubt that there are differences in behavior between individuals. This is obvious from observing the mating strategies of individual males during the breeding season. But when we tried to determine paternity using DNA fingerprinting, it could not be done because we could not distinguish one potential father from another. This procedure, used for determining paternity in most mammals, as well as southern elephant seals, did not work with northern elephant seals because they lacked genetic diversity. Too bad because it would have been important to determine whether the males that dominated mating fathered most of the pups, or whether dominant males that mated frequently suffered a decrease in sperm count and ability to inseminate females, or whether males that mated exclusively with departing females (who had already mated with other males) or forced copulations fathered offspring. Questions such as these could not be answered. Just how important is the loss of genetic variability of elephant seals, or any other animal, is not clear. It is unlikely that a population ever regains the range of genetic diversity it once possessed. Lastly, our observations of behavioral differences in animals with genetic similarities suggest a flexible link between genes and behavior.

Many other mammals have extremely low genetic variation because their populations underwent bottlenecks. European bison living today descend from 12 individuals (Luenser et al. 2005). The giant Panda, the golden snub-nosed monkey, the golden hamster, and cheetahs are others whose populations experienced bottlenecks (O'Brien et al. 1985; Caro 1994; Zhang et al. 2002). Among birds, the New Zealand black robin and greater prairie chickens are genetically depauperate for the same reason. The Galapagos Islands giant tortoise is a prime example of an animal that suffered a population bottleneck (Milinkovitch et al. 2013). But the classic example of the bottleneck phenomenon, and recovery from it, is the northern elephant seal.

2 Back from the Abyss, Population Recovery, and Genetic Aftermath

If you are going through hell, keep going.

Winston Churchill

The entire northern elephant seal population that exists today was reconstituted from the small remnant herd that survived the 1890s on Isla de Guadalupe, Mexico. The herd persisted, in large part, because the island is remote in the open ocean, about 247 km due west of mid-Baja California, Mexico. It is basically a mountaintop in deep ocean waters, fully exposed to the long sweep, strong winds, currents, and swells of the Pacific Ocean (Figure 2.1). One has to have a good reason to go there because it is not on the direct route to anywhere else, and it is difficult and costly to access. Once there, it is dangerous to land a dinghy, especially on the exposed western side of the island, which is like being on the prow of a ship exposed head-on to the full force of the ocean's wrath.

This small remnant herd, contrary to all expectations, endured and, with time, recovered from near extinction. R. C. Banks coined the term "extation" to describe "the status of a species whose population has been reduced to such a low level that it can no longer function as a significant part of its normal ecosystem ... or to the point where there is considerable doubt whether the species remains extant" (Banks 1976; Le Boeuf 1977). It is a less cumbersome term than conceptions such as nearly extinct, probably extinct, and brought to the brink of extinction. The northern elephant seal is a prime example of an animal that has come back from extation. The resurgence of the population is the most remarkable, well-documented, comeback from near annihilation ever recorded. In this day and age when hundreds of species go extinct daily, including millions of birds in the last three decades, there are few feel-good stories such as this.

How did the recovery unfold? What made it possible? The recovery was slow, uncertain, and there were setbacks early on. In 1907, just a few years after the population's low point, Charles Miller Harris came upon a small herd. He promptly shot 10 seals for the Rothschild Museum. Habits die hard. Clearly, there was still no guarantee that the animals would survive. In 1922, a joint Mexican–American expedition to Guadalupe counted 264 seals. Concluding that the seals were almost certainly going to be exploited again, the Mexican government gave complete protection to the species. Legislation was not sufficient; a small garrison of soldiers was

Figure 2.1 Three elephant seal colonies. (a) Pilot Rock Beach on the northeast point of Isla de Guadalupe, Mexico. The beach is crowded with elephant seals at the peak of the breeding season. (b) A view of Elephant Seal Beach on the northwest side of Isla de Guadalupe. Access to this major breeding beach is difficult. (c) Año Nuevo Island and mainland. (d) The breeding area for elephant seals at Piedras Blancas in southern California. Courtesy of Friends of the Elephant Seals. (A black and white version of this figure will appear in some formats. For the color version, please refer to the plate section.)

stationed on the island to give teeth to the prohibition to kill or capture elephant seals. When the seals began to appear in southern California waters a few years later, the United States gave the species legal sanctuary as well.

The number of seals began to increase slowly during the early twentieth century, and in time, this rate of increase gained considerable momentum (Bartholomew and Hubbs 1960; Radford et al. 1965; Le Boeuf et al. 1974; Allen et al. 1989; Stewart et al. 1994; Lowry et al. 2014, 2020). The seals began immigrating to the north from Isla de Guadalupe and breeding on other Mexican islands to the east near the mainland, such as Islas San Benito and Los Coronados, and then on southern California islands such as San Miguel, Santa Barbara, and San Nicolas between 1925 and 1949 (Figure 2.2). The population was estimated at 13,000 in 1957. In the

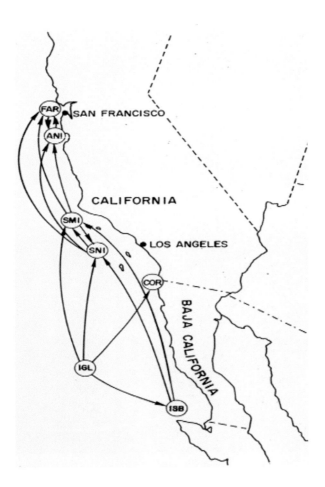

Figure 2.2 The dispersion of elephant seals during the late 1960s and the 1970s from Isla de Guadalupe (IGL) to Islas San Benito (ISB) and Los Coronados (COR), from IGL and ISB to southern California islands such as San Miguel Island (SMI) and San Nicolas Island (SNI), and from SMI and SNI to central California breeding sites such as Año Nuevo Island (ANI) and the Farallones (FAR).

1960s and 1970s, the seals began colonizing breeding sites in central California, such as Año Nuevo Island, the Farallones, and Point Reyes.

By the 1970s, colony numbers on the Mexican colonies had stabilized, and San Miguel Island in southern California had become the largest colony. In 1976, the population had increased to 48,000 seals. In the decades that followed, immigrants from San Miguel, which was getting crowded, established new colonies on adjacent Channel Islands, such as Santa Rosa, and mainland sites such as Piedras Blancas near Cambria and San Simeon, California.

Over the course of a little more than 100 years, the northern elephant seal population increased from an estimated 20 seals in 1894 to about 239,000 in 2010. The population increased at an annual rate of 3.8% from 1988 to 2010. It has continued to increase, judging by the last census in southern California in 2013 (Lowry et al. 2020). It is estimated that 250,000–300,000 seals exist in 2020. The majority of pups, about 90% of them, are produced in the Channel Islands in southern California. Evidently, the seals have recovered in number, and they now breed in what is thought to have been their former breeding range (Figure 2.3).

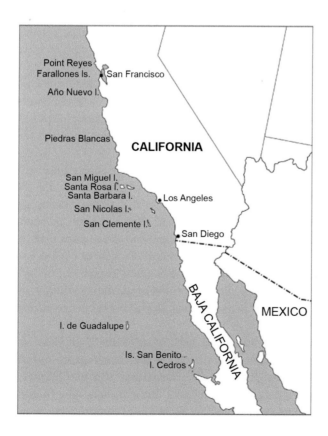

Figure 2.3 The major colonies of northern elephant seals in 2020.

In contrast, the numbers in Mexico have decreased since the early 1990s at two of the three largest rookeries, Isla de Guadalupe and Islas San Benito, while the Isla Cedros numbers have remained stable (Garcia-Aguilar et al. 2018). The decline in population in Mexico since the 1990s is positively correlated with rising sea surface temperatures and associated increasing air temperatures, linked with global warming. Heat is a problem for seals breeding on land, given their thick blubber coats that serve them in cold waters; their metabolic rate increases, and they risk hyperthermia and heat stroke (Hansen and Lavigne 1997). Moreover, pups are especially vulnerable because of their high mass-specific metabolism (Codde et al. 2016). During an extremely warm winter in the late 2000s, pup mortality on one San Benito island was greater than 90% (Salogni et al. 2015). J. P. Gallo (pers. comm.) reports that the sand on breeding beaches is decreasing owing to hurricanes. The data from Mexico suggest that the population's breeding range is moving north to colder waters.

What are the factors that facilitated the population recovery? Obviously, government protection and enforcement of laws preventing killing or capture were important, especially early in the recovery when the license to kill was virtually considered a God-given right. Later, in 1972, the Marine Mammal Protection Act prohibited taking all marine mammals and enacted a moratorium on the import, export, and sale of any marine mammals, or parts of them, within the United States. "Taking" included hunting, killing, capture, and/or harassment of any kind. Second, as the population increased in the early stages of recovery, the seals chose to breed on islands that were isolated and relatively uninhabited by humans and terrestrial predators. Third, since elephant seals feed in deep waters on fishes that human fishers cannot reach or catch, or consider low priority "junk fish," they avoided direct conflict with the human fishing industry. This is important because the habitat necessary for them to feed and flourish remained open and available during the population recovery. Loss of habitat is the primary cause of extinction of most species.

Today, there are few obvious threats to the continued existence of the northern elephant seal population. The big exception is climate change and the associated warming waters that cause redistribution of prey that makes foraging more difficult. This is a huge problem that affects us all and bears watching and actions to curtail climate change. Another is the conflict between seals and humans over beach use that may limit breeding area expansion. This is likely to occur at Piedras Blancas, where seals now frequent or occupy 12.9 km of coastline (Figure 2.1). The site borders Highway 1 and is accessible to the large human population in the Los Angeles area. What will happen when seals use a beach to breed and humans demand the same beach to recreate?

There are 12 major rookeries in existence today (Figure 2.3). Three are in Mexico: Isla de Guadalupe, Islas San Benito, and Isla Cedros. Nine are in California: San Clemente, Santa Barbara, San Miguel, Santa Rosa, and San Nicolas in southern California, a mainland rookery at Piedras Blancas, and Año Nuevo Island and mainland, the Farallon Islands, and the Point Reyes peninsula in central California. It is unlikely that major new colonies will be established north of Point Reyes, California, because suitable island or mainland breeding sites are lacking along the west coast until one reaches Puget Sound, in Washington state, and Vancouver Island, Canada. Puget Sound is a large shipping port inhabited by millions of people in the Seattle area, and Vancouver Island has few suitable beaches for breeding and supports a large industrial fishing operation, which is incompatible with the presence and movements of the seals. Will the total population continue to increase and, if so, to what degree and in what locations? Population expansion is expected to continue but at a reduced rate because pup mortality increases as crowding increases. Sandy beaches, the preferred habitat of the seals for breeding, are almost totally occupied at San Miguel Island and San Nicolas Island, and this will happen soon at nearby Santa Rosa Island. The number of beaches at other islands in southern California is limited. There is room for expansion of breeding at Point Reyes and Piedras Blancas, and this is already happening at the latter site. Other mainland sites are available but human presence may prevent colony establishment.

The largest northern elephant seal rookery in 2010 was San Miguel Island (16,208 births), followed by San Nicolas Island (10,882 births), Santa Rosa Island (5,946 births), and Piedras Blancas (4,469 births). Of these four, Santa Rosa Island exhibited the highest average annual growth rate (45.6%). The large San Miguel colony has become crowded and is leveling off in number (Lowry et al. 2020). Other colonies that have decreased in number, besides the Mexican colonies noted above, are Santa Barbara Island, Cape San Martin/Gorda, South Farallon Island, and Año Nuevo Island and mainland in California.

Growth at the mainland site near Piedras Blancas has been explosive since breeding began here in 1992. The average annual rate of increase was a whopping 8.1% from 1992 to 1995, which is an exponential growth. This period was followed by a more modest rate of increase after 1996, which is typical of the entire population. The colony continues to grow. In 2015, more than 23,000 elephant seals were counted along a six-mile stretch of the coastline north and south of Piedras Blancas. By 2018, the total number of seals had increased to 25,520 along an eight-mile stretch of the coast (Figure 2.1). The number of pups produced increased from 4,469 in 2010 to 5,800 in 2018. It is notable that elephant seals have learned to exploit mainland sites to breed.

It is necessary to give special attention to the growth and development of the Año Nuevo rookery because it has been subjected to intensive study. Most of what we know about these animals comes from studies conducted at this rookery. More than 200 peer-reviewed research articles have been published from studies of elephant seals at this location. By Año Nuevo, we refer to both the island and the adjacent peninsula on the mainland where seals breed (Figure 2.1). A channel, approximately 400 m wide, separates the island from the mainland. Close monitoring of this colony reveals how a colony develops and changes over time and, by extrapolation, the dynamics of the entire population, as well as the factors that underlie fluctuations in number (Le Boeuf et al. 2011).

From the population perspective, the Año Nuevo rookery is a rather small peripheral elephant seal breeding site in central California. It has, however, been inhabited by pinnipeds (seals and sea lions) for a long time. In 1584, Spanish explorer Francisco de Gali, when returning from the Philippines, sailed in close enough to note the numerous seals that abounded in the area (Le Boeuf 1981).

No elephant seals were seen on the island from 1890 to 1948 by the US Coast Guard personnel based on the island to maintain a lighthouse or by the California Fish and Game deputies that visited the island periodically to census sea lions. In July 1955, two subadult male elephant seals were observed on the island, and in subsequent years, non-breeding seals continued to be observed in increasing numbers. In 1961, elephant seals began breeding on the island with the birth of two pups (Radford et al. 1965).

After the first births, the number of pups produced on the island increased slowly over the years. By the 1968 breeding season, when University of California researchers began their long-term studies of the seals, 180 pups were produced. During the early 1970s, the two breeding beaches on the island began getting crowded. Young females from Año Nuevo Island migrated to form new colonies at Southeast Farallon Island in 1972 and crossed the channel and began breeding on the Año Nuevo mainland in 1975 (Le Boeuf and Panken 1977). A total of 750 pups were produced at both sites in 1977.

Pup production on the mainland increased rapidly, while that on the island decreased. Crowding on the island was associated with pup mortality, and this prompted females to move to the mainland, where they had the freedom to space out and therefore pup mortality was minimal. In 1987, more pups were produced on the mainland than on the island; the mainland became the principal breeding site in the area. Pup production on the mainland peaked in 1995 at 2,041 pups. From 1995 to 2005, pup production on the mainland stabilized at about 2,000, but fell to 1,725 in 2010 and rose again to about 2,000 in 2019 (Figure 2.4).

What do these fluctuations in numbers tell us? First, the seals increase in number until it becomes crowded during the breeding season. This makes it difficult for young females, especially those giving birth for the first time, to wean their pups. They lose out in competition with older, larger females and are relegated to the periphery of harems where they are pestered by males and are most likely to get separated from their pups. The high mortality they incur prompts them to seek out new, less crowded places to give birth. They emigrate and form new colonies before the peak of crowding. They are the vanguard for creating new colonies, forced to emigrate by reproductive persecution. The Año Nuevo colony was created by females from San Miguel Island seeking less crowded conditions to breed (Figure 2.2). Colonies at the Farallones and Point Reyes were formed by females from Año Nuevo seeking better and safer breeding opportunities. This is how new colonies are created and how the population expands: crowding increases pup mortality that causes females to disperse (Le Boeuf et al. 2011).

What causes fluctuation in colony numbers and pup production? Basically, this is due to external or internal recruitment of breeders. For example, colonies in central California – Año Nuevo, the Farallones, and Point Reyes – grew rapidly because of the influx of breeding-age females from large colonies in southern California. In 1971 and 1972, 43% of the males and females in residence during the breeding season

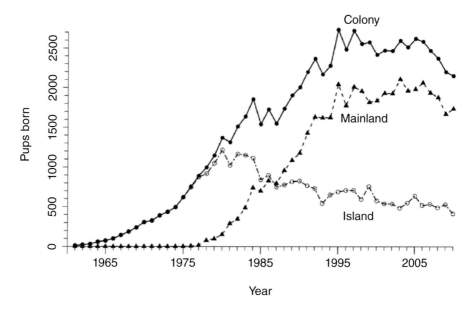

Figure 2.4 Pup production at the Año Nuevo colony from 1961 to 2010. From Le Boeuf et al. (2011).

at Año Nuevo were born in southern California. Two decades later, 1989–1998, the percentage of breeding females at Año Nuevo that were immigrants from San Miguel Island and San Nicolas Island had declined to 30%, and the immigration rate was further reduced to 13% in 1999 through 2003 and then to 6% in 2006 through 2009. The Año Nuevo rookery stopped growing when the immigration rate from southern California decreased. Why did the immigration rate decrease? The cessation of growth and decline of pups born at Año Nuevo since 1995 was coincident with the explosive growth of new colonies at Santa Rosa Island and Piedras Blancas on the adjacent mainland. These colonies were created by females moving from San Miguel Island, which had become crowded. This was an easy move because these sites were close by. In effect, the seals stopped migrating to central California because suitable breeding sites were available nearby. Why not? The consequence was a decrease in growth at Año Nuevo to such an extent that the colony would not have increased in number from internal recruitment alone.

The take-home message is that the process we have described of southern California rookeries sending immigrants to form new colonies in central California is integral to the population's dynamics over time. Immigrants from Mexican colonies created new colonies in southern California, and immigrants from southern California formed new colonies further north. Density dependence and pup mortality are the drivers. Of course, there are other factors that operate as well, such as global warming, high sea surface temperatures, and vulnerability to weather at the new site, which might impact weaning rate, and there is the matter of the survival of the weaned pups at the new site. We know that weanlings born at Año Nuevo and the Farallones suffer high mortality during the first foraging trips at sea (Chapter 7). One would expect weanling survival to be higher at a mother colony such as San Miguel Island, but we don't know this for sure because this study has not been conducted.

So, the seals made a remarkable comeback from extation and appear to be thriving today. Is the population similar to what it was 130 years ago? Does any population of animals that has been bludgeoned to the virtual brink of extinction fully recover? Can we simply view the recovery in numbers and breeding range as reflecting a successful comeback from near extinction and feel good about it? Well, certainly not from a genetic perspective. The present population is genetically depauperate; it lacks genetic variability. A practical indication of this is that one cannot use DNA fingerprinting to determine paternity in this species; that is, one cannot use a technique that works with most other animals to determine the father of a pup. It makes sense that genetic variability was lost when the population was reduced to a couple of dozen individuals. Experiencing one or more population bottlenecks not only devastates the population but also leads to the loss of many genes. Consequently, we assume that the population today has less genetic variability than the population that existed before sealing began in 1800. We assume that genetic variability increases adaptability; it is the best hedge against an uncertain and constantly changing environment. Is this the case? We don't know and it is difficult to test. We do know that it takes a long time to recover genetic variability that is lost and the recovery depends on the mutation rate and the success of the mutants.

3   The Year of the Seal

Being asked what animal you'd like to be is a trick question; you're already an animal.

Doug Coupland

If you want to view wild elephant seals in nature, to best plan your visit you will want to know what the animals are doing. This is the objective of this chapter.

Some background on human interest and response to large animal aggregations provides context. Historically, animals that gathered in large numbers were exploited by humans for food or profit. In the sixteenth century, seabird rookeries were pillaged for meat, eggs, feathers for down or quill pens, and guano for fertilizer. A century or two later, shore-breeding marine mammals were killed for food, ivory, fur, and oil. In the nineteenth and the early part of the twentieth centuries, the scale of exploitation was so great that many animal populations were decimated.

In recent times, the focus has shifted from harvesting wildlife to viewing animal aggregations for pleasure, entertainment, or knowledge. Popular wildlife attractions are large terrestrial African mammals, whales, dolphins, and colonial nesting birds (Hoyt 1993; Rivarola et al. 2001; Yorio et al. 2001; Okello et al. 2008). The cultural and economic benefits of wildlife watching are immense, and such activities have increased exponentially in the last few decades (Hoyt 1992; Tapper 2006). But there is also a downside: too much tourism disturbs the animals being viewed. Numerous studies support this statement. Whale-watching is a case in point. With nine million people in 87 countries and territories watching cetaceans on an annual basis, sustainability of the approach is difficult: "There are not enough accessible friendly cetaceans with time available – after feeding and socializing – to allow for nine million close encounters per year" (Hoyt 2003). Restricting viewer access is necessary.

In any case, when large groups of animals are accessible, stationary, and predictable in space and time, thousands of people may observe thousands of animals on a daily basis at a single site. Viewing spectacles of this kind represents the extreme end of the dimension of wildlife watching. Those that are accessible to the public are immensely popular, raise revenue, contribute to livelihoods, create awareness, and often support conservation. Such is the case today with elephant seal breeding colonies on the mainland of the west coast of the United States, which began in 1975 (Le Boeuf and Panken 1977; Le Boeuf and Campagna 2013).

The two accessible mainland breeding colonies in the world for northern elephant seals are located at Año Nuevo State Park (Figure 3.1) and Piedras Blancas (Figure 3.2), part of Hearst San Simeon State Park in California. Thousands of elephant seals inhabit these two sites year-round. Southern elephant seals breed on a single mainland site at Península Valdés, Patagonia, Argentina, which welcomes visitors (Le Boeuf and Campagna 2013).

Año Nuevo State Park is located 34.6 km north of Santa Cruz and 92.5 km south of San Francisco and is reached by Highway 1. The park's Natural Preserve designation, managed by Park Rangers, has the objective of protecting the elephant seals along with many other local animals such as otters, sea lions, coyotes, marine birds, native plants, and the intertidal ecosystem (Le Boeuf and Kaza 1981). Elephant seals can be observed here up close year-round either on a docent-led tour during the breeding season or through a self-guided permit system during the non-breeding season. During the elephant seal breeding season, December through March, one must be on a guided walk, limited to 25 people, to view the seals (Figure 3.3). Up to 25 tours are conducted per day. Reservations are made through www.reservecalifornia.com or the Año Nuevo State Park Website. Parking is available at the park's entrance.

Volunteer naturalists with deep knowledge of the seals lead the tours and keep the elephant seals and humans safe from each other. The seals are viewed from a distance of 20 m or more, sufficiently distant to avoid affecting their behavior. Elephant seal viewing requires a moderate hike over varied terrain, including sand dunes, which takes about 2½ hours. Sturdy shoes, layered clothing, water, and rain/wind/sun protection are recommended. Visitors requiring mobility assistance can make a reservation for an Equal Access Tour. In the non-breeding season, April through November, visitor permits to see the seals on your own inside the Natural Preserve are issued from 8:30 a.m. to 3:30 p.m.

The seal colony at Piedras Blancas is located near Hearst Castle at San Simeon in central California (Figure 3.3). The site is managed by Friends of the Elephant Seal, a non-profit organization of volunteers dedicated to educating people about elephant seals and other marine life and to teaching stewardship for the ocean off the central coast of California. See http://www.elephantseal.org. The organization cooperates with California State Parks.

The Piedras Blancas elephant seal rookery spreads over about 10 km of shoreline around Piedras Blancas on the central coast of California. The viewing areas are located 145 km south of Monterey, 8 km north of Hearst Castle State Historical Monument in San Simeon, 2.4 km south of Piedras Blancas, and a few hours' drive up Highway 101 from Los Angeles County, the most populous county in the United States. The viewing areas are open every day of the year, are wheelchair accessible, and free (Figure 3.4). Reservations are not required. Parking is a short walk from the principal viewing area. In 2019, approximately 800,000 people visited this site to view the seals.

You can just drop in unannounced at Piedras Blancas, if you can find a place to park. For comparison, it is a lot more accessible, less costly, and more exciting to view elephant seals breeding here than to visit Africa to view sleeping lions or one or two

(a)

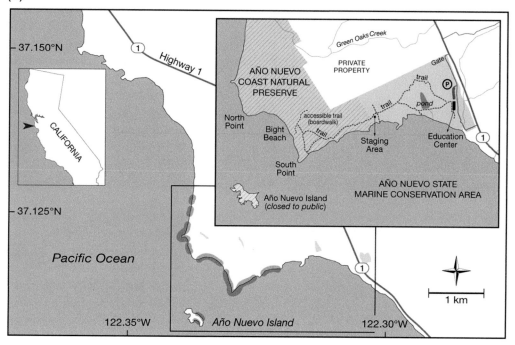

(b)

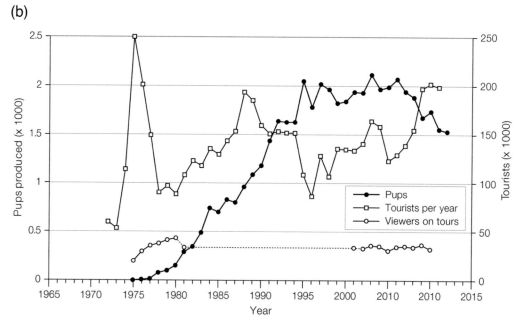

Figure 3.1 The Año Nuevo colony showing (a) the distribution of elephant seals at the peak of the breeding season (dark areas) and (b) pups produced and visitors. Adapted from Le Boeuf and Campagna (2013).

(a)

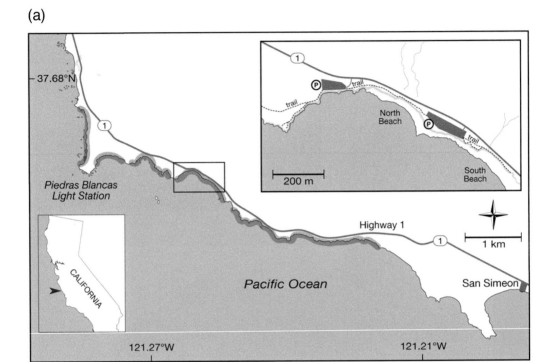

(b)

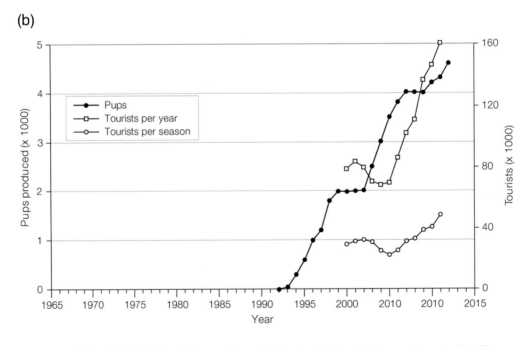

Figure 3.2 The Piedras Blancas colony site showing (a) the viewing area, boardwalk trail, parking lots (P), and the elephant seal distribution at the peak of the breeding season (dark areas) and (b) number of pups produced and visitors per year in contact with docents during the breeding season. Adapted from Le Boeuf and Campagna (2013).

(a)

(b)

Figure 3.3 Places to view elephant seals. (a) Visitors on a guided tour view the elephant seals at Año Nuevo from a distance. (b) A view of a portion of the elephant seal rookery at Piedras Blancas during the breeding season. The perspective is viewing south; Highway 1 is to the left of the photograph. (A black and white version of this figure will appear in some formats. For the color version, please refer to the plate section.)

cheetahs resting on a tree. During the breeding season, the seals are a challenge to watch because they are so active; it is like watching a three-ring circus (Box 3.1).

In summary, there are substantial differences between the two viewing sites. Año Nuevo is a walking park exposed to the sun and winter wind and rain; it is accessible to many, but it takes some effort. It provides a special wildlife experience because the guide-to-viewer ratio is high; the interpreter is a source of information with deep knowledge of the seals and the area; the viewers are on nearly the same level as the seals, in close proximity to harems; and elephant seals may be moving on all sides of the small group of viewers, which keeps the viewers alert and makes some guides nervous. Piedras Blancas is a drop-in park where large groups of people can view the elephant seals quickly, for free, with little effort, and visitors can linger all day if they wish. It is close to a large metropolitan population. The view from the boardwalk at Piedras Blancas is close to the elephant seals, panoramic, and safe, but the viewer may be in a crowd of other viewers, and the guide-to-viewer ratio is low, giving less

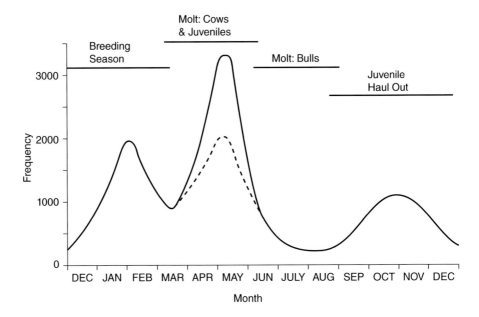

Figure 3.4 The annual cycle of northern elephant seals divided into four categories. The number of seals censused during the molt of cows and juveniles varies from dotted line to the solid line. Adapted from Le Boeuf and Laws (1994).

Box 3.1 The Rutowski Reflex

Observe elephant seals sleeping on the beach, no matter what their age or sex. If the animal is sleeping on its left side, it breathes through its nostril on the right side only (the nostril furthest from the substrate), keeping its left nostril tightly shut. If the animal turns over to its right side, the opposite holds; it breathes through the left nostril only and keeps the right nostril clamped down. This reflex is named after its discoverer, Ronald Rutowski, an astute undergraduate observer way back in the 1960s. Ron was a budding biologist and musician at the time, who went on to have a long and successful career studying butterflies.

The function of the Rutowski reflex? Obviously, the reflex reduces inhaling sand. Selection for diving adaptations under high pressure may have enabled the reflex. That is, nasal anatomy evolved to allow the nostrils to be clamped down independently. I've often wondered about how finely tuned the response to body position relative to the substrate might be and its sensory basis. One could gauge the sensitivity of the response if a seal was restrained and rotated slowly, as if on a rotisserie, through 360°. Does the reflex vary with the substrate, for example, sand or mud? Scientists will spend hours pondering such things. What other animals exhibit this reflex? Now, there is a PhD thesis.

opportunity to learn from the interpreters the value of the unique wildlife being viewed. Piedras Blancas provides an incredible view of many elephant seals up close for the viewer in a hurry.

Quarterly Categories of Events during the Year

The annual cycle of events in the year of elephant seals described here is based on observations at the Año Nuevo rookery. The pattern holds for Piedras Blancas as well, except that events at the southern colony may occur a day or two earlier than at the northern colony.

The elephant seal calendar starts on December 1 and ends on November 30. It is useful to divide the annual cycle into four terrestrial phases: breeding season, female and juvenile molt, male molt, and juvenile haul-out (Figure 3.4).

Breeding Season: December to Mid-March

The breeding season begins in early December with the arrival of adult males (Figure 3.5). Most of the major competitors are on the rookery by December 25. Concurrent with the arrival of adult males is a rapid decline in the number of juveniles,

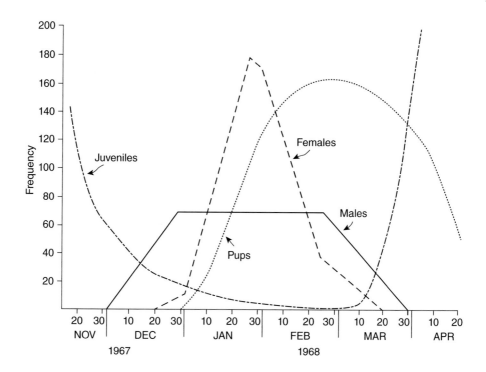

Figure 3.5 An approximation of the relative frequency of juveniles, pups, females, and males during the breeding season.

one to four years old, that previously were abundant on the rookery in October and November. Pregnant females begin arriving in early to mid-December. Their numbers reach a peak during the period between January 26 and February 2 and then decrease until all of them have returned to sea by about March 10, leaving their newly weaned pups on the rookery. Young males begin leaving the rookery in late February, but the larger bulls remain there until the end of March, a week or two after the last female has departed.

The pupping period is from about the third week of December to the end of the first week of February. Copulations occur from the first week of January through the first week of March, with February 14 being the peak day of copulation frequency. Valentine's day! Is this sheer coincidence or magic?

The breeding season is the best time to view the seals if you want to see a lot of activity and a variety of behaviors. The number of animals present is at its peak in late January when the rookery is a circus of activity. You are likely to see males competing with each other, threatening and fighting for dominance, possibly a bloody fight between high-ranking males that lasts 15 minutes or more. Females will be seen arriving pregnant, giving birth, nursing their pups, squabbling with neighboring females over space, copulating, and negotiating their way through a gauntlet of suitors as they attempt to return to sea after weaning their pups. You will see suckling pups, orphaned pups, dead pups, and weanlings being chased out of harems. You will see weaned pups experiencing the water for the first time and learning to swim and dive. The background noise adds to the urgency at hand: males bellowing threats to other males, females vociferously rejecting male advances and females emitting pup attraction calls to their pups, the hungry pups crying to get their mother to expose her nipples, and the forlorn wailing of weanlings whose mothers have abandoned them. Overhead, gulls squawk and wheel and compete for the nutritious placentas disgorged after females give birth. Indeed, the squabbling, shrieking gulls are a sure sign that a birth has just occurred.

Of course, these animals are dangerous, and you are on their turf, so be wary. Males will run over you as if you were a piece of furniture in their way. Don't take it personally. Females are protective of their pups and will deliver a vicious bite if you come too close. Be careful where you walk – sleeping males in the dunes can look like large immobile rocks – and be keenly aware of what is behind you at all times and have an escape route in mind at all times. Have fun.

Female and Juvenile Molt

There are two attractions for visitors during the period from mid-March to the end of May, the weaned pups and the molting females and juveniles. The weaned pups left behind by their mothers will be seen sleeping in the dunes in groups or pods during the day and entering the water tentatively at dusk. You will see them splashing about, learning to swim and dive. As they age, they become more proficient and spend more time in the water until reaching a point where they are in the water throughout the night and sleeping in the dunes during the day. Most of them will depart on their first foraging trip in unknown waters by the end of April.

As the last females wean their pups and return to sea, the first females that gave birth early in December begin returning from sea in mid-March. They return to the rookery to molt, a process that takes about one month to complete. The influx of adult females continues for about two months. The adult females are joined by juveniles of both sexes that are one to four years old. The highest number of seals in a year is seen on the rookery in late April. Still present in the early spring but declining in number are adult males; all of them return to sea by the end of March. You will also see large numbers of weanlings, the pups of the year. Approximately 80% of them leave the rookery by the end of April. Rather suddenly, in early to late May, there is a rapid decline in the number of seals on the rookery. This mass departure to sea is coincident with a marine upwelling of cold water off the California coast that increases the availability of nutrients that support growth of plankton, which, in turn, provides food for fish, marine mammals, and birds.

At this time of the year, you will see many seals on the beach in large groups. Most of them will be sleeping, especially on warm, sunny days. The seals will be relatively peaceful; there is nothing to compete for except possibly space to rest and sleep. After foraging at sea for about 2½ months, females return to the rookery to molt (Figure 3.6). Molting involves the replacement of all hair and epidermis (Worthy et al. 1992). Elephant seals molt once a year; they do this on land and the process takes on average 32 days. During molting, seals fast and do not go into the water unless it is exceptionally warm. During the process, they lose 25% of their starting mass at the rate of 3 kg per day.

Figure 3.6 A subadult male in mid-molt.

Molting starts around the eyes, ears, anus, and penile openings and then proceeds along the ventral midline and from both the anterior and posterior dorsal midline areas. The molt then spreads over the body surface, eventually replacing the entire pelage. It is a drastic process unlike the shedding process in many terrestrial animals, such as dogs. In elephant seals and their relatives, the monk seals, the hair is shed along with large sheets of cornified epidermis through which the hairs penetrate. This continuous sheet of keratin breaks loose from the underlying tissue. You will see a carpet of sloughed off hair and skin on the beach or blowing around in the wind.

In the 1960s, when residents were not accustomed to seeing elephant seals in the area, because they were just starting to recolonize central California, an elderly person saw a molting elephant seal on the beach near her house. She was concerned because it looked awful. She called the police to investigate. The policeman agreed that the seal looked miserable so he shot it to put it out of its misery. Of course, the seal was fine; it was just molting, as residents came to learn with the increasing influx of seals in the following years.

After molting, the seals look reborn and brand-new. All freshly molted elephant seals are uniformly blackish grey or silver, and the hairs are initially very short but they double in length of about 3–4 mm by the completion of the molt. Thus, most new hair and the underlying epidermis grow while the seals are on the beach, before they return to sea.

Pups are born with black fur, called lanugo. They molt their lanugo shortly after weaning at four to six weeks of age. This is the most obvious change in pelage. The black natal coast is replaced by a shiny silver coat. Elephant seals never look more charming, if that is the right word, than a newly molted weanling.

Male Molt: June through August

The lowest number of animals on the rookery is observed in June, July, and August, when breeding-age males molt. Young pubertal males, five to six years old, tend to arrive in early summer, and they are followed by the older males in late summer. A few young of the year and 1 ½-year-olds are seen early in this period, but they make up less than 5% of the total seals in residence. Juvenile numbers begin to increase in August, and in some years, they outnumber molting males by the end of the month.

The seals group together in large clumps on the beach. Males that fought viciously in January may be found lying together in close contact. Indeed, the gonads of males regress and do not produce sperm and androgens during the off season. Males do not fight except for young males that may engage in play fighting or practice fighting. Molting proceeds as described for adult females.

Juvenile Haul-Out: September through November

By the first week of September, all but a few males have returned to sea to forage and fatten up for the approaching breeding season. Young of the year and juveniles up to four years of age increase steadily in number to a peak in October. At Año Nuevo, many of these juveniles are immigrants from southern rookeries. Although molting does not usually occur during this period, some yearlings exhibit a pathological skin and pelage condition described as "scabby molt." Scabby molt is a skin disease observed in some yearlings (less than 5%) as they appear on the rookery at this time.

Pubertal subadult males begin to arrive on the rookery during the middle of November, a time when juveniles are decreasing in number.

The events described here summarize elephant seal life on land. The other half of their dual existence is their life at sea, where they spend the majority of their time. Of course, you cannot see them at this time and you only know they are at sea because they are absent on land. Many aspects of their behavior at sea, however, are well known from measuring instruments carried by the seals. These are described in Chapter 9, on diving, foraging, and migration.

4   Fieldwork 101

Getting There and Getting Started

Do not go where the path may lead, go instead
where there is no path and leave a trail.

                                                            Ralph Waldo Emerson

Simply getting access to the animals for a field study is a big deal. It may be costly, difficult to arrange, and unpredictable. The study site for some animals may be distant and remote, requiring permits that take time to snake through the glacially slow bureaucratic process, dangerous because of the political situation, and uncomfortably cold or hot or infested with biting insects. The conditions vary widely, and different field studies pose different problems; imagine the difference between conducting research in Africa, the Antarctic, and the tropics. Getting there is the first obstacle and the modes of travel range widely: commercial airline, helicopter, automobile, all-terrain vehicle, boat or launch, snowmobile, swim, travel on foot, and so on. For remote sites, one has to purchase food and water, package it, and set up shelter, in addition to packing all of the instruments necessary to conduct the research.

What Can Be Done

A major obstacle is the animals themselves. Having arrived where the animals live, their behavior determines what data you can collect and how you do it. Is the animal group relatively stationary or do the animals move about over several kilometers per day? Imagine collecting data on wandering Cape hunting dogs or spotted hyenas or African lions, which must be tracked in all-terrain vehicles, or imagine tracking mountain sheep on foot, as opposed to sitting in place to observe a creche of emperor penguins or a rookery of California sea lions. It takes time and energy to find and follow animals that roam over extensive ranges. Equally important is the question of whether the animals are shy and flee if they see you or whether they are unafraid and unmoved by your presence. In the former case, one is limited by what can be done. For example, if sea lions or fur seals are frightened, they all rush *en masse* into the water, even females in the act of giving birth; it takes hours before they return to place and calm down. In contrast, some animals are oblivious to your presence; the sea lions that inhabit the Galapagos Islands can be approached to within a body length. It is fundamental that the researcher does not affect the behavior of the animal under study.

Finally, there is the matter of approachability and manipulability. Animals that live and breed in a stationary place, and only move about in this restricted place, that are not frightened of you, and that are approachable and can be manipulated for certain operations such as tagging, blood collection, or weighing provide the best opportunities for collecting data bearing on important biological principles. Elephant seals are ideal subjects in this regard. When I was a graduate student, the white rat was the ideal subject for research in the laboratory; the elephant seal is the field equivalent of the white rat. An ideal subject is one that permits you to address important biological questions easily and efficiently.

Of course there are other obstacles in field work, and they are endless. The tribulations range from the annoying, such as being defecated on in bat caves, to being dive-bombed from breeding western gulls, to getting parasitized by bloodsucking leeches while studying amphibians up to your neck in a lake at night, to having to wait for a year or more to get chimps to tolerate your presence, as Jane Goodall learned. Some animals are downright dangerous to study. When Ian Stirling was studying polar bears in the Canadian Arctic in the 1970s, he found it prudent to carry a high-powered rifle (Stirling 1975). Even today, researchers are advised to carry flares, bangers, pepper spray, and firearms and to travel in groups, if possible. When setting up a camp near bears, researchers string a trip-wire alarm system or an electric fence around the campsite or set a watch.

Getting to Año Nuevo Island

Compared to the field studies of most of my colleagues, I had it easy. Getting to the elephant seal rookery at Año Nuevo Island from my office and lab at the university took less than a half hour by car. That was the easy part. Once there, however, we had to launch a craft on the beach to cross the treacherous channel to get to the island. In the early years, we sometimes felt like Odysseus having to survive Scylla and Charybdis to get home. The special problem of studying pinnipeds has long been having to navigate to the island rookeries they inhabit, and when the weather is inclement this can be unpredictable, difficult, or dangerous.

When I began conducting research on elephant seals at Año Nuevo Island in December 1967, the sole breeding site was on the island, about a kilometer from the mainland. The island is 30 km north of Santa Cruz, California, at 37° 06. 5′ N latitude, 122° 20.2′ W longitude, and projects out into the Pacific. It consists of 12 acres of shifting sand and layered beds of Miocene cherty shale slanting into the sea. About 100 years earlier, it was connected to the land, but the sea's constant battering carved out an island from the peninsula. It is home to California sea lions, Steller sea lions, harbor seals, and even a few sea otters and the nesting place for western gulls, pigeon guillemots, and black oystercatchers, and other shorebirds. For a while, it was the universe to 70 brush rabbits whose ancestors were brought here by a local prankster in 1949. Año Nuevo Island was the first piece of land sighted by Sebastian Vizcaino on the New Year of 1603. It is thought to have been a former workshop for making arrow heads and other artifacts by the indigenous people that lived here before the Europeans arrived. This place was part of growing up on the north coast to several generations of Steeles, McCrarys, and McLeans. It is a good place to fish and collect abalones.

The low rocky profile of the Año Nuevo peninsula has long been a danger to shipping and was the site of numerous shipwrecks during the nineteenth century (Le Boeuf 1981). This prompted the federal government to purchase the island in 1870. A foghorn was installed in 1872, then a lighthouse in 1890. A large Victorian house was constructed for the keepers in 1904 but the facility was difficult to maintain, dangerous, and expensive. Numerous tragedies and deaths occurred to the lighthouse keepers crossing the channel from the island to the mainland. The facility was abandoned in 1948 and replaced by an automated buoy. The buildings and support structures on the island were left to go to ruin. In the early twentieth century, the dangerous channel continued taking its toll on boaters, for it was the scene of numerous drownings, rescues, boat crashes, and stranded skin divers. For example, a small boat overturned in the channel in mid-July 1957, and five people drowned; four people had lost their lives in this channel the previous week (*San Francisco Chronicle*, July 22, 1957).

In 1958, the State of California purchased the island and part of the adjacent mainland from the federal government. A temporary permit was granted to the Stanford Research Institute (SRI) to study the resident seal and sea lion populations from 1961 to 1966. The chief scientists were Tom Poulter from SRI and Robert Orr from the California Academy of Science. After elephant seals were observed breeding on the island in 1959, they instituted a variety of studies of the pinnipeds on the island during the period 1961–1966. In addition to conducting censuses, they tagged elephant seal pups and yearlings and branded adults during each breeding season, resulting in approximately 500 marked animals (Orr and Poulter 1965, 1967; Radford et al. 1965). There were plans to construct a laboratory on site to further research efforts, but this did not pan out. Poulter and Orr relinquished their permit, and their research on the island ended. Their tagging and branding efforts, however, gave us a head start on our tagging program that followed and continues to this day.

On June 29, 1967, Año Nuevo Island was declared off limits to the public. On July 1, 1968, the University of California at Santa Cruz was granted a lease to conduct research on the island fauna, largely through the efforts of my colleague Richard Peterson, a newly hired assistant professor of biology who specialized in the behavior of sea lions and fur seals. The face of the island didn't change much when the university took over as "concessioner." The buildings continued to fall apart. The major positive for us researchers was that the Coast Guard gasoline storage building, a cement blockhouse with walls 30 cm thick, was intact and indestructible. It was outfitted with a stove, refrigerator, and lights – all run on propane – cabinets, a desk, and three sleeping bunks. An outdoor privy was constructed nearby. Having shelter and a place to cook and sleep enabled us to stay overnight as well as to spend long weekends on the island rather than having to cross the channel every day. This was a boon to us and allowed us to monitor our subjects from dawn to dusk, as well as on moonlit nights (Box 4.1).

> **Box 4.1** There Is a Seal in the House
>
> Two undergraduate students and I were turning in, exhausted after a long day of tagging seals. Our sleeping quarters were in the old abandoned Foghorn House on Año Nuevo Island. The conditions were minimal, sleeping bags on military-type bunks with hard springs, damp mattresses, a concrete floor, and no heat. It was midwinter and the peak of the breeding season for the elephant seals. As usual, the bulls were amped up and stampeding all over the island, fighting and threatening each other with their loud guttural vocalizations. This cacophony went on all night long. Not exactly guaranteed to induce sleep like *Eine Kleine Nachtmusik* or Bach's *Goldberg Variations*. Just after midnight the front door slammed open with a bang. Dazed and half asleep, we fumbled for flashlights and aimed the light at the front door to see a nearly full-grown bull backing caterpillar-like into the bunkhouse, being pushed and forced backward into it by a more dominant aggressive male. The three of us sat bolt upright, eyes wide open and in shock. (I see this as a *New Yorker* cartoon wanting a label.) We fumbled to light a couple of Coleman lanterns and assessed the situation, which was at the same time scary, annoying, dangerous, and humorous. We put clothes on as quickly as if the place were on fire. Shouting across the room, we agreed that getting the seal out of the bunkhouse was necessary or we faced the even more dangerous alternative of sleeping outside. How? The seal barely fitted through the door jambs and this was the only exit, short of punching a hole through the wall, which we didn't have the tools to do. After a couple of frustrating and scary hours of alternatingly trying to maneuver the seal and dodging it, we yelled and probed with sticks and lifted arms extending pillows and tried to appear taller than the seal (which sometimes causes a male to back down); we managed to get the interloper to shuffle back out by squeezing out through the door backward. We slammed the door shut and barricaded it with unused bunks. Whew! It was a long night. We laughed about it the next morning.

Here, I recount some of the problems and stories of getting to the island. In the late 1960s and the following decade, this was, at times, fun, easy, and adventurous and at other times unpredictable, scary, and dangerous. It depended on the weather, wind, fog, treacherous currents, and surf conditions; all of these elements varied with time of year. The channel separating the island from the mainland continued to be a perilous passage that had to be negotiated carefully. Five- to six-foot whitecaps slanted in from the north, breaking near mid-channel and washing clear across to the other side where they crashed against the smaller and faster southern waves, creating a string of rippling rooster tails that shot up 10 feet or more in the air. The middle of the channel was shallow at low tides, with hidden rocks that had to be avoided. A rusting pipe, a former marker, broke the surface and provided a useful reference point if you could see it through the fog, surf, and spray. The trick was to face the incoming waves from the north, power through the crest, and after the wave broke, veer quickly to the left to avoid the next wave, and power through mid-channel and into the lee of the island. This had to be done without hitting rocks, shearing a pin, and losing motor power.

Losing power meant drifting into the hazardous big stuff, and capsizing. The crossing could be challenging, and experience and local knowledge were vital for providing a cognitive map of what was possible under the prevailing conditions.

It took two or more of us to launch a boat through the surf on the mainland. Wet suits provided protection from the cold waters and flotation, if needed. We went through several crafts to learn what worked best: a Boston whaler with a 25 hp Johnson, an inflatable Avon rubber raft with a 5 hp Johnson, and eventually a more robust Zodiac rubber raft with a hard bottom and a 15 hp Johnson. The Boston whaler was too heavy, difficult to launch, and took several hands to haul it out of the surf zone when landing. Its flat prow made it difficult to face waves before they broke and invited disaster. We scuttled the whaler and replaced it with a 12-foot-long Avon inflatable powered by a 5 hp Johnson. This was one of the early inflatable designs with a soft bottom that slumped or caterpillared over waves rather than punching through them. Its biggest drawback was that it was too slow to get out of trouble and it could not accommodate a heavier, more powerful outboard motor. The ideal raft for the crossing turned out to be a hard-bottom Zodiac inflatable powered by a 15 hp Johnson. It was fast enough and could be launched and landed by two people. The motor could be removed from the craft and one person could carry it. The Zodiac performed well in facing the breaking surf, and its three separate floatation compartments kept the vessel afloat even if one compartment was punctured.

We had some memorable crossings. The strategy for a typical winter crossing was like plotting the strategy to par a tough golf hole created by a master course designer. The prevailing wind, the surf, the currents, and the shallow mid-channel with rocks nearly exposed were the hazards one had to consider in getting the raft and its occupants across the channel to a safe landing on the island. Positioning and orienting the raft correctly at all times with respect to the oncoming surf was critical because having to correct an error increased the odds of taking on water, losing power, or capsizing. Even after negotiating the dangerous mid-channel, the homestretch was still a problem because of the many seals and hundreds of sea lions lounging in the shallows offshore or scurrying away in fright. I remember one crossing in particular. As we throttled down the outboard motor and prepared to jump over the sides to pull the raft ashore through the shallows, a bull elephant seal reared up to about eight feet in front of us and dropped the full weight of his head and forequarters with teeth bared into the front compartment of the rubber raft. The craft lost pressure as fast as a balloon when a pin is stuck in it. Fortunately, we were close enough to shore to lift the raft up onto the beach, while avoiding further contact with the belligerent seal. We learned a lesson: don't take elephant seals lightly – males or females – when you are in a rubber raft.

In the winter of 1968, Dick Peterson and I drove up the coast intending to launch the slow Avon inflatable to visit the island. But damn! The nefarious coastal fog was so thick we could not see the island. We could barely make out the mild glow of the sun overhead through the thick pea soup fog. It looked like we would have to abort our plan, but Dick, ever the optimist, said, "We can make it." I gave in, reluctantly, to his supreme confidence. We launched and put-putted away from the mainland, keeping an eye on the sun somewhere overhead. Otherwise, we could not see beyond the prow of the raft, but we listened for the raucous racket of sea lions vocalizing and waves breaking to get a sense of where we were. After about 20 min, we breathed a sigh of relief as we heard waves breaking on the shore and we coasted in for a landing on the sandy beach. But to our amazement, we had landed only several yards from the mainland shore where we had launched. We had come full circle!

Part of doing research on this island is that we were often asked to take visitors (Box 4.2). It was good public relations to take important people, and the administration

---

Box 4.2 The Head-Hunting Professor

We had just crossed the channel and were going down to the seal beach to begin the day's work. Joining the expedition today was an anatomy professor from a neighboring university. He was a head-hunter. He collected seal skulls, as well as other post-cranial bones, for use in his research. Elephant seal skulls were especially prized, and rare in collections, because elephant seals rarely die on land or wash up onshore; most die at sea. So, the professor was especially delighted to encounter a moribund adult male elephant seal as we arrived on the beach. He expressed his delight, set his kit on the beach near the seal, and began sharpening his dissection knives. Ready to start, he kicked the seal with his boot, as a would-be car buyer often does, and perhaps to get a sense of the condition of the carcass. The seal reared its head up, roared, and turned about to face the professor who had sprung back and fallen on his back in a state of shock. Moral: sleeping seals resemble dead seals or rocks or boulders.

Figure B4.2.1 Elephant seals can resemble dead seals, smooth rocks, or boulders, or even mountains. (A black and white version of this figure will appear in some formats. For the color version, please refer to the plate section.)

encouraged this. I remember a famous professor of literature from the University of California at Berkeley who wanted to see the island on a Saturday in January. Larry, a strapping undergraduate, and I had gone to the island the day before and we had arranged to meet him on the mainland launching site Saturday morning at about 11 a. m. But when we woke up on Saturday morning it was clear that the channel crossing was going to be hazardous. The wind was howling, and the surf was up, and white waves were breaking in mid-channel. We would not have attempted a crossing under these conditions, but we had no way of communicating with the professor to abort so we climbed into our wet wetsuits reluctantly and steeled ourselves for a cold, bumpy, teeth-clenching ride. We crashed in on a wave on the mainland beach. Our manner – forced to be sure – was that this was normal and typical for the time of year. We told the professor, who had no wetsuit, to sit on the floor of the raft, which already held about six inches of cold water, and we pushed the raft through the shore break and started the motor. When we got to mid-channel we hit a rock, sheared the pin, and lost power. I told Larry to jump over the side and hold the raft in place to keep us from drifting while I replaced the broken shear pin. It was low tide, and the cold water came up to his neck as he struggled to hold the craft against the strong current. "This happens," I said as casually as possible. "Don't be alarmed," as I dealt with the shear pin with ice-cold fingers. By now, the professor was soaking wet and shivering. The pin replaced, I started the motor. Larry hauled himself in over the side, and we continued to the island. We changed into dry clothing, gulped a glass of red wine – a most welcome tradition after a nasty crossing. We gave the professor a quick tour of the island and had a bite to eat. The condition of the sea and the channel had moderated considerably on our return in mid-afternoon and the crossing was uneventful. We returned to Santa Cruz, changed again into dry clothes, and were welcomed by the Chancellor, at his house on campus, for wine and cheese. The professor was exuberant and commented on how much he had enjoyed the adventure and loved every minute of it. Sure, I mumbled to myself.

And there was the time we had arranged to meet a famous Hollywood actor and his girlfriend who wanted to see the seals. At the appointed time, we met them on the mainland beach. It was raining, and the wind was so strong it was blowing sand horizontally. She was dressed in a fur coat and wore 3- or 4-inch heels; he was dressed for a walk on Rodeo Drive. Not right. We had lunch in nearby Davenport instead.

In the mid-1970s, the Director of Parks and Recreation at Año Nuevo State Reserve, Roger Werts, wanted to check out his domain, which included visiting the island. My graduate student, Michael Bonnell, and a couple of undergraduates had arranged to pilot the raft and make the crossing. It didn't go well. The raft lost power in mid-channel, and the current and surf took them quickly south through the turkey tails; the raft overturned and all they could do was hang on as they were at the mercy of the currents and drifted south. Roger could not swim. I got a call from Lud McCrary at Big Creek Lumber, who had a view of the island, telling me of the accident. I raced up the coast by car and pulled over at Waddell Creek, a few miles south of the island, just in time to witness the last stage of the rescue operation. I photographed the last passenger being air lifted from the raft by a hovering helicopter. The sea was raging, the waves were 6–10 feet high, the wind was blowing spray from the white caps. All passengers survived. Whew! Roger Werts did not see the island. I don't think he ever did. Later in the day, the raft washed through the surf on Waddell Creek beach. The motor was gone, and the raft was crumpled, deflated, and demolished.

The Seals Change Breeding Location

By 1975, the island was starting to get crowded with elephant seals and a few of them began breeding on the mainland. In their December 1973 issue, *Sunset* magazine announced this unusual spectacle to the public. It recommended the 30-minute walk to Año Nuevo point to see the animals. This changed everything. Suddenly, the dunes and beaches were flooded with hundreds and then thousands of people eager to see the seals. People beat a swath to the beach, flattened the delicate plant life on the sand dunes, milled and circled around each seal, threw rocks at sleeping bulls to get action photographs, and blithely walked within biting distance of a bull's head, ignorant of the danger. We on the island observed the transformation of the mainland point through binoculars, in awe of the seal–human confrontations, alternately afraid for the people and then for the seals. The Department of Parks and Recreation had a monster on its hands. Something had to be done immediately. It was too expensive to close and police the beaches, and the few part-time rangers were insufficient to act as guides. In discussions with the head park rangers, we decided that the best solution was to train university students to act as guides to lead the visitors on tours. We were most familiar with the seals and their approachability so we set up the rules for how close the tourists could get to the animals. We taught classes on the Año Nuevo area, its flora and fauna, and, of course, the elephant seals and wrote a book that was used in classes (Le Boeuf and Kaza 1981). Later, the guided tours became the responsibility of volunteers, which worked out quite well and saved the state a lot of money.

The important thing for us researchers is that, in a short time, the mainland supplanted the island as the seal's central breeding area (Box 4.3). This meant that we no longer had to negotiate the dangerous channel to the island to study the seals, and most of our research efforts since then have been done on the mainland where about 90% of the elephant seals reside. We thanked the seals for their consideration.

Box 4.3 Nature Calls on a Dark Night
With males roaming all over the top of the island at night fighting with each other, going to the outdoor toilet was a severe test of bravery for novice field workers. On dark moonless nights, it was the last resort, because you could only see what the flashlight lit up but could hear the galumphing sounds of males moving urgently to the left and right and behind you. Moreover, they have excellent night vision and can see you. This was no place for old men with urgent bladders.

Figure 4.1 Censusing elephant seals at Pilot Rock Beach on Isla de Guadalupe, Mexico. The censuser is seated on the rock outcropping in the lower left of the photograph.

Getting to Other Islands and the Surprises Encountered

But enough about the local scene; we had to get to many other islands where seals and sea lions bred during these early years. Censusing throughout the breeding range of the elephant seals was necessary to document population status and growth, as well as colony health, movement between colonies, and the colonization process, and to tag seals to monitor their movements, growth and development, and reproductive success (Figure 4.1). We made multiple expeditions to various islands where elephant seals bred during the period 1968–1978. These rookeries included Isla de Guadalupe, Islas San Benito, Isla Cedros, Isla San Martín, and Los Coronados in Mexico as well as San Miguel, San Nicolas, Santa Rosa, Santa Cruz, San Clemente, and Santa Barbara Islands in southern California. These were the principal rookeries in the entire breeding range of elephant seals at the time.

Visits to these islands not only enabled us to collect the data and meet our objectives, but being in these remote sites with little or no human traffic, we were in a privileged situation. We observed and encountered unexpected events that were of such importance that they had to be dealt with immediately. Some of them turned into long-term projects. Here are some examples.

A New Colony of Fur Seals

On one of our first expeditions to San Miguel Island during July 20–21, 1968, we reached the exposed western tip of the island after a four- hour trip from Oxnard on the mainland on a vessel provided by the National Park Service. We offloaded a Boston whaler to prepare to go ashore at Adam's Cove. Yes, this is the same heavy, unwieldy launch with the flat bow that we didn't like in crossing the channel to Año Nuevo Island. The surf was kicking up good, and we had no alternative but to time our landing between waves. We followed a big one, gunned the motor, rode the wave in, and crash-landed on the beach in the middle of a pod of northern fur seals, *Callorhinus ursinus*. What! This was as surprising as coming face to face with an alien vehicle from outer space. What were they doing here? This animal, also called the Alaska fur seal, bred nowhere else in the world at the time except on the Pribilofs and a few other islands in the Bering and Okhotsk Seas. But this was clearly a breeding unit. The colony consisted of approximately 100 individuals, including an adult male, about 60 females and 40 young ones in natal pelage, that is, newborn pups. Five adult females bore tags or tag scars, showing that they were born on American or Russian Islands in the Bering Sea. We tagged 36 pups. Within a month, we published an article on the discovery in *Nature* magazine (Peterson et al. 1968).

The saga did not end well for us. Fur seals had long been exploited for their fur, and a treaty had been signed by four countries to divide the profits: the United States, Canada, Russia, and Japan. The US government had oversight over the "resource," being responsible for the annual killing on the Pribilofs as well as the status of the "stock." So, it was not surprising that federal agents from the National Marine Fisheries Service (NMFS) visited the colony on San Miguel Island and essentially told us, "Thank you very much. We'll take over from here." They closed down the area to all visitors and prohibited us from conducting follow-up studies or visiting the area for studying other seals. Dick was distraught because these animals were his major research interest; he had studied these animals for his PhD thesis.

Since this time, NMFS agents have counted the fur seals annually. The colony has flourished (Melin et al. 2008). The "stock" was estimated at 14,050 in 2016. Research on the seals and sea lions and other animal and bird life on San Miguel Island was shut down, and the place remains closed to researchers in 2020. This closure is to protect the fur seals that can no longer be killed for their furs.

Oily Seals

Dick Peterson and I were on San Miguel Island tagging weaned elephant seals shortly after crude oil from the Santa Barbara Channel oil spill washed up on Northwest Cove on Point Bennett on March 17, 1969. Union Oil Platform A had blown out on January 28, 1969, the largest oil spill in the United States at the time, and media coverage commanded the front page of newspapers from the moment the oil reached the shore. Now, the petroleum sludge had washed up on the distant, pristine beaches of the western most point of the island, an especially rich area for seals, sea lions, and marine birds, and the newspapers announced that hundreds of seals were dying. A wealthy person from Los Angeles proposed to send out a barge to clean and rescue the animals. Walter Hickel, the director of the Department of the Interior was under fire.

Northwest Cove was drenched in black, stinky, sticky crude oil, a foot or more deep in places, and the silver pelage of 100 or so newly molted, weaned pups in the cove

Figure 4.2 Santa Barbara Channel oil spill on March 17, 1969. (a) Northwest Cove on the western tip of Point Bennett, San Miguel Island, showing the crude oil that washed ashore in mid-March 2017 and blackened beaches and soiled weaned elephant seal pups. (b) A close-up view of weaned elephant seals enmired in crude oil at Northwest Cove. (c) Dick Peterson attaches a tag to the hind flippers of a weaned elephant seal pup with 75% of its body covered in crude oil. (A black and white version of this figure will appear in some formats. For the color version, please refer to the plate section.)

were blackened and drenched with it (Figure 4.2). What was the effect of the crude oil on the seals? Were they endangered? Did it kill them? We were in a position to answer these questions.

On March 25, we attached yellow-colored, numbered plastic or monel tags to the interdigital webbing of the hind flippers of 58 weanlings and 5 yearlings that were at least 75% covered with a mixture of crude oil, sand, and detritus. (In some pups, this included the entire head.) An equal number of clean pups were tagged in the adjacent cove area to provide a control group for comparison. The crude oil was gooey, tenacious stuff that stuck like glue to our hands, and except for the good fortune of finding a washed up 50-gallon drum holding a bit of gasoline, we would have had to live with sticky fingers until we returned to the mainland. Our boots were beyond recovery and we threw them away. I didn't have much hope that the oily seals would recover and survive.

Sightings of the tagged seals 1–15 months later revealed that survival was higher in the oily seal group (40%) than the clean seal group (31%)! This difference was upheld in a follow-up of the tagged seals years later (up to 1976); 9.5% of the experimental group were seen again compared to 7.9% of the control group. All re-sighted animals in each group were in apparent good health, and the pelage of all the seals in the oily group showed no trace of oil just a month after tagging. How did they get rid of the oil? Eight seals from each group had swum 408 km to Año Nuevo Island, indicating that they were sufficiently healthy to swim a long distance.

The Santa Barbara Channel oil spill was a disaster. It certainly killed thousands of birds. This was not the case for the elephant seals on San Miguel Island. The weaned pups were fasting at the time and did not ingest the oil, which merely adhered to their fur and skin. We published a report stating that the crude oil that coated many weaned elephant seals in March 1969 had no significant immediate or long-term deleterious effects on their health (Le Boeuf 1971b). This was not what some reporters, who were keen to curtail oil drilling in the Southern California Bight, wanted to hear. I understand this political stance, but a scientist has no choice but to report the facts. In June 1969, a reporter and a photographer from *Life* magazine rejected the idea that the large sea mammals were not negatively affected by the oil, because they counted over 100 dead seals and sea lions along a stretch of beach black with oil. Of course, given the abundance of animals at this location, one can count a 100 or more dead animals on any day of the year; after all, they die from various causes.

Premature Sea Lion Pups

During our visits to the Channel Islands in southern California and the offshore islands in Baja California, Mexico, in the Spring of 1968, 1969, and 1970, we observed hundreds of moribund, premature California sea lion pups. They littered the beaches in places such as San Miguel Island and San Nicolas Island in southern California and on Islas San Benito, Cedros Island, and Natividad in Mexico (Le Boeuf and Bonnell 1971). In some cases, a female protected the dead fetus and took it with her as she entered the water.

Premature pupping could represent the normal variation in the population, or the problem may have been human-generated. Exposure to high levels of DDT, especially near the superfund site at San Pedro in southern California, where the Montrose Chemical Corporation deposited tons of DDE over the years, was suspected because these organochlorine pesticides can mimic estrogen and induce abortions.

The case was made that premature births were associated with high organochlorine pollutant residue levels (DeLong et al. 1973). For example, 242 dead premature pups were counted on San Miguel Island on April 25, 1970, and 348 on May 18, 1971; on San Nicolas Island 442 premature pups were reported between January 17 and May 3, 1970. Organochlorine pesticides and polychlorinated biphenyl residues were two to eight times higher in tissues of premature parturient females and pups than in similar tissues of full-term females and pups on San Miguel Island collected in 1970.

It is still not clear why the premature birth rate was so high in the late 1960s and early 1970s. In subsequent years, the rate decreased precipitously, and this was coincident with the cessation of DDT manufacturing and dumping of byproducts into the sea in about 1972.

The association between high levels of DDT residues and premature pupping remains unclear; many confounding factors are in play (O'Shea and Brownell 1998). The sea lion population has more than doubled since the early 1970s. One thing is clear, however; considerable research on the high levels of organochlorine pesticides in pinnipeds has been conducted over the last several decades. We did a lot of work on this important problem as well (Le Boeuf and Bonnell 1971; Lieberg-Clark et al. 1995; Le Boeuf 2002; Le Boeuf et al. 2002; Kannan et al. 2004; Debier et al. 2005a, 2005b, 2006). The jury is still out on the effect of organochlorine pollutants on pinnipeds.

Dialects in Seals

Traipsing across island beaches in southern California and Mexico in 1968 to census and tag seals, with bulls issuing their threat vocalization throughout the day and night, it was obvious to us that the males on different islands "spoke a different language." To be more precise, this sounded like vocal dialects to us, consistent differences in the predominant, stereotyped, threat vocalizations of bulls from different islands. At this time, dialects were known only in humans and some birds. Elephant seals were added to the list. We present our findings in Chapter 12.

The lesson we learned about getting out to the field to conduct research and what can be done depends greatly on the traits of the animal under study. Some animals have habits and demeanor that make them easier to study than others. Elephant seals are an ideal study animal in this regard and this is why we know so much about them. Another lesson is that being "out there" exposes one to the unexpected. This is a good thing and often takes one's planned research in a different direction.

Marking Individuals

Having arrived at the study site, one must decide what to do first. To understand what is going on when one is observing animals in nature, identifying individuals is vitally important. It is first base; you cannot get very far without doing this. Lyndon B. Johnson said, "In a nation of millions and a world of billions, the individual is still the first and basic agent of change." He was talking about humans, but the statement applies to all life. Natural selection acts on individuals.

Recognizing individuals and being able to follow them daily, monthly, and from year to year is necessary for understanding the behavior of animals, just as numbers or names on jerseys help us understand the football game we are watching. This was the first problem that we faced when we began studying elephant seals in the late 1960s. We could envision the great scientific potential of studying elephant seals, but exploiting this potential depended on being able to identify individuals.

To mark and identify animals, however, you must get close to them, and we were naïve about these huge animals with their wolf-like canines. We didn't know if this could be done and how dangerous it might be. After all, some bulls weigh over two tons and are aggressive, and our few initial observations showed that they could move fast over short distances. The females, although considerably smaller at about 500 kg (1,102 lb), were even more dangerous because they had pups to protect and they did this with vicious resolve.

Getting a mark, any kind of mark, on a seal was our top priority. Clearly, it seemed most likely and least dangerous when the animal was isolated from others and sleeping, which the seals are inclined to do on warm, sunny days. It wasn't just fear of the animal to be marked but the other males and females in the vicinity that might sneak up on you while you were concentrating on applying the mark. One had to be attentive to the surroundings and have an exit strategy. It was very important not to trip or fall on the first few steps of your get-away.

Dick's idea was to fill a fire extinguisher with black dye, sneak up to a sleeping male, and direct a stream of the dye onto its back. We flipped a coin and he lost. He pumped up the fire extinguisher, approached the prey cautiously, and pressed the button. The spray, under pressure, struck the male on the back and he bolted upright in apparent rage. Dick tumbled back head over heels, the fire extinguisher pinwheeling over his head, and he scurried away to safety. Failed experiment.

Our next attempt was to put colored paint in a plastic sandwich bag, seal it, and hurl it with sufficient force at a seal to break it, leaving a splat of paint on the animal. Then, we described the mark and named the seal; for example, red left neck became RLN. The marks lasted the length of the breeding season, which enabled us to record the social and reproductive behavior of the marked individuals. However, after "painting" 10–20 seals, things got messy. Firing paint balls from a paint pellet pistol offered some improvement especially when conditions made it impossible to approach the target closely. Eventually, we settled on a solution that served our purpose for years and still does (Figure 4.3). By now we had learned that we could approach a sleeping seal and we could write an identifying mark or name on its back or side without risking life and limb. We used a plastic squeeze bottle filled with one of two solutions. The first was to bleach the pelage using a mixture of Lady Clairol Ultra Blue and hydrogen peroxide. We were gifted bottles of Lady Clairol from the company over the years and suggested the slogan "the beauty and the beast, they both use Lady Clairol." The bleach worked best on sunny days and the yellow-bleach hair marks could be read easily from a distance throughout the breeding season. The second solution was a black dye that worked equally well. This low-tech method was a critical breakthrough in understanding the game these animals played during the breeding season. Now we could track the behavior of individuals.

Elephant seals molt during the summer, however, and their entire pelage sloughs off and the identifying marks disappear. To recognize individuals from year to year, and throughout their lives, we needed a more permanent mark. We inserted cattle ear tags – colored, numbered plastic tags – in the interdigital webbing of one or both hind flippers. At the beginning of the next breeding season, we read the tag and remarked the individual with bleach or dye, giving it the same it had the year before. Thus, each marked individual had a name and a tag number and, for the record, a card showing all the vital information such as date of tagging and location.

Figure 4.3 Marking seals. (a) Marking a sleeping adult male with a bleaching solution. (b) Marking an adult female in a crowd of females. (c) A mother and her pup marked with identical paint spots. (d) A subadult male, Irv, marked with the bleaching solution. (A black and white version of this figure will appear in some formats. For the color version, please refer to the plate section.)

Tagging weaned pups at about five weeks of age was easy and straightforward. At weaning, the mother leaves her pup and departs from the harem to go to sea and feed. The weanlings exit from the harems and spend most of the time sleeping in the sand dunes nearby. To study development and determine migration and inter-rookery movements, we tagged thousands of weanlings at Año Nuevo, as well as at various other rookeries along the Pacific coast of the United States and Mexico, especially during the late 1960s and 1970s. Pups at each rookery were given a different colored tag, blue for Isla de Guadalupe, Mexico, yellow for San Miguel Island, red for San Nicolas Island, green for Año Nuevo Island, pink for the Farallones, and so on.

Tagging adult females and males was more difficult. It required a strong and firm hand, more caution, and quick reflexes to escape when the seal sprang up as the tag was applied. In subsequent years, we conducted physiological studies that required sedating seals in the field. One experiment required drugging mothers two days after giving birth as well as just before they weaned their pups (Ortiz et al. 1984; Costa et al. 1986). Another required transporting weaned pups to the laboratory for procedures and then returning them to the beach where they were captured. What was the effect of these procedures on the animals? By all criteria we used – weight loss, time and reliability of returning to the same site, subsequent survival, and reproductive success – the experimentals did not differ from control animals (Le Boeuf 1995). These animals are robust against disturbance.

In summary, marking and tagging individuals was critically important in helping us to understand the natural history of this wild animal. Marking males revealed the mating strategies of males and the lifetime mating success of individual males and females. The rangers leading tours at Año Nuevo State Reserve can convey a wealth of information to the tourists on the natural history of these animals because of the research that has been conducted here by University of California researchers tracking individual seals for over 50 years. The story would be far less informative if individuals had not been monitored daily throughout each breeding season and throughout their lives.

5   Adapting to Life at Sea and on Land

Just as we have two eyes and two feet, duality is part of life.

Carlos Santana

The adaptation to two vastly different environments, land and sea, is a remarkable characteristic of all pinnipeds. Elephant seals challenge the limits in both environments. Twice a year the elephant seals feed at sea and twice a year they breed, molt, or rest on land (Figure 5.1). Adult males spend eight months of the year at sea, and adult females spend ten months of the year at sea. Both sexes are submerged 90% of the time they are in the marine environment (Le Boeuf et al. 1988, 1989, 2000a; Le Boeuf 1994). The simple math reveals that males spend 60% and females 75% of the year underwater. These air-breathing mammals spend more time holding their breath underwater than breathing in air. As Kramer (1988) noted, "Should we call them divers or surfacers?" What we call them is somewhat arbitrary. It is clear, however, that they live in two different worlds, the sea and the land. We would be justified in calling them either marine mammals or terrestrial mammals.

The origin of the pinnipeds is both obscure and controversial (Riedman 1990). Phocids are thought to have evolved from an otter-like carnivore during the Miocene about 20–25 million years ago. One thing, however, is indisputable: the ancestors of pinnipeds were originally land mammals; they began entering the ocean to feed on abundant food resources. We see something like the beginning of this transitional process in modern times; on some South Pacific islands, pigs swim, wade, dive, and graze over coral reefs in search of food at low tide, and along the coast of Scotland, domestic sheep swim to graze on algae in the intertidal zone. It takes time for land animals to become efficient at exploiting the sea for food, and this requires significant adaptations.

Life at Sea

What changes or adaptations are required for a mammal to make a living in the ocean? Kooyman (1989) addresses three categories: physiological, anatomical (energetics), and behavioral. We summarize some of the key adaptations.

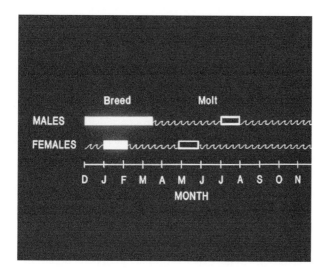

Figure 5.1 The months of breeding and molting and time at sea for adults of both sexes.

Physiological

The first hurdle a foraging mammal in water faces is that it cannot breathe; it must hold its breath and tolerate anoxia, the absence of oxygen. Elsner and Gooden (1983, p. 1) emphasize the vital importance of oxygen deprivation that diving mammals face:

Depriving living organisms of oxygen and of the opportunity to rid themselves of carbon dioxide leads inevitably to progressive disruption of cellular biochemical processes, physiological dysfunction and the undoing of biological integrations which are collectively essential for homeostatic life. This condition, asphyxia, is a threat which lurks never far away from living vertebrate organisms.

How does the elephant seal deal with this fundamental problem? To pursue and capture prey, the seal must have sufficient oxygen to stay underwater until it finds the prey; this usually takes about 20 minutes but may take as long as 2 hours. As Kooyman (1989) points out, oxygen is the most critical resource for aerobic and prolonged natural diving. It underwrites the diving system, and there has been strong selection pressure for its enhancement and concentration. The following are the most important ways that divers, like elephant seals, increase their oxygen stores: (1) reduce oxygen utilization by restricting blood flow to certain organs; (2) increase myoglobin concentration; (3) tolerate arterial oxygen tension to levels so low that it would render a human unconscious; (4) increase blood oxygen capacity that results in high concentrations of hemoglobin and red blood cells; and (5) support an exceptionally large blood volume.

First of all, elephant seals dive with a high oxygen load. The lungs are not the principal reservoir for oxygen, as is the case with humans and most other animals. Blood is their most important storage site of oxygen, containing 65% of oxygen stores, followed by muscle (30%) and the lungs (5%) (Kooyman 1985, 1989). The body of adult females is 21.2% blood, which is high compared to terrestrial mammals; by the time juvenile elephant seals are ready to go to sea for the first time, their body is 14.4% blood (Thorson and Le Boeuf 1994). In line with these priorities, elephant seals exhale before diving. This not only reduces oxygen in the lungs but also reduces the inspiration of nitrogen (the lungs are a better storage site for nitrogen than oxygen), which can cause high blood nitrogen tensions. Moreover, the seal's lungs collapse at 25–50 m, removing oxygen from the lungs and alveoli and sending it into nonabsorptive pockets in the upper body that are less prone to pressure (Falke et al. 1985). The flexible trachea collapses as well. The lungs collapse on every dive and reinflate on return to the surface! The fact that these diving seals are not discomfited by lung collapse is a remarkable adaptation. Lung collapse is life threatening for humans.

When an adult elephant seal dives, it shunts oxygenated blood to vital organs like the heart, brain, and muscles at the expense of other peripheral organs that go without oxygen or with less oxygen. Its heart beat decreases from about 80 to 90 beats per minute at the surface to 30 to 34 beats per minute during dives (Andrews et al. 1997, 2000; Hindell and Lea 1998; Le Boeuf et al. 2000b) but the bradycardia can decrease to as low as 4 beats per minute. Upon hearing this, one of my colleagues said: "This is like diving while dead." This means that the seal can reduce the amount of oxygen used during the course of a dive. In addition, metabolic depression is facilitated by continuous exposure to cold waters and ingestion of cold prey. The diving seal can lower its temperature by about 5°C. Turning down the furnace means using less oxygen and this helps the seal prolong its dives.

The seal has other physiological tricks for managing its oxygen stores. Elephant seals have a high hemoglobin concentration as well as a high myoglobin concentration. Indeed, myoglobin concentrations are 10–30 times greater in aquatic species than terrestrial species. High oxygen concentrations are coupled with a high tolerance to low oxygen tension, which is especially evident in their kidneys and brains. Elephant seals can sequester red blood cells in reservoirs such as the hepatic sinus and spleen and modulate oxygenated blood into the vascular system as needed while diving (Thornton et al. 1997a, 1997b, 2001, 2005; for further details see Box 9.1 in Chapter 9, MRI Scans in Elephant Seals!).

Kooyman (1989) argued that it was important to predict the limit of duration of aerobic dives as this would be important in understanding the behavior of diving animals and their ecology, especially the manner in which they partition and manage oxygen stores. He defined the aerobic dive limit (ADL) as the maximum breath-hold that is possible without any increase in blood lactic acid (LA) concentration during or after the dive. Calculation of the ADL is difficult, and estimates vary even for the same species. In general, when a diver exceeds its ADL, some organs may have to rely on anaerobic metabolism and LA accumulates, and it must spend time at the surface following a dive getting rid of LA before it can dive again. For example, if the dive of a Weddell seal exceeds 17 minutes, it requires more time at the surface before it can dive again.

The ADL has received much attention in the literature on diving animals. The diving behavior of elephant seals, however, does not fit well with expectations. In an early study (Le Boeuf et al. 1988), adult females exceeded the predicted ADL of 17 minutes 37% of the time. Additional studies showed that even after dives lasting an hour or more, the seal spent only two to three minutes at the surface before diving again, suggesting that there was no need to get rid of LA at the surface because anaerobic metabolism was not used. It is possible that the seal slept during some of these dives, hence lowering its metabolism and increasing its ADL. In any case, the elephant seal stands out, in contrast with most other diving mammals, with regard to adhering to a calculated ADL. This segues into a notable point: to the extent that metabolic depression is characteristic of diving in elephant seals, diving might also serve multiple behaviors besides feeding such as sleep, conservation of energy, and predator avoidance. After all, the seals spend less energy diving than resting at the surface! Recent studies of the behavior of the diving seals in three dimensions suggests that considerable time is spent drifting, which may indicate sleep or rest (Mitani et al. 2010).

How is predation decreased by rest or sleep while submerged? Seals at the surface are vulnerable to predation by near-surface predators like white sharks and killer whales. The ocean surface is where they hunt. Seals that sleep below the surface are in a safer place. Being below the predator, they are less likely to be observed or attacked. For example, killer whales prey on southern elephant seals at Marion Island. The seals sleep or rest below the depths that the whales dive, which is 300 m (Aguilar de Soto et al. 2020). Thus, minimizing the time at the sea surface is adaptive (see also Le Boeuf et al. 2000a). Moreover, it is natural for seals to sleep submerged while holding their breath. This is what they do on land in what we call terrestrial dives – sleep while breath-holding for 10–25 minutes and be awake for 2–3 minutes when breathing (see Chapter 11). Thus, if the seal must sleep, it is safest to do it at depth. Moreover, besides predator avoidance, sleeping at depth may be coupled with digestion after a prolonged bout of foraging.

In summary, increases in blood volume, hemoglobin concentration, and myoglobin concentration during development result in a large oxygen storage capacity, which, combined with an increased ability to decrease metabolic rate while diving, enables elephant seals to maximize time underwater for their business, which may be travel, foraging, rest, sleep, or predator avoidance. The elephant seal's diving strategy is to make haste slowly, *festina lente*. While at sea, they take in as much energy as possible foraging while expending as little energy as possible. Isn't natural selection wonderful? This is referred to as the paradox of diving.

Compression

To descend to 500, 1,000, or 1,500 m, the seal must withstand incredible pressure and great changes in pressure. This is perhaps the most formidable challenge for a mammal making a living in the ocean. Because the pressure increases by one additional atmosphere for each 10.06 m descended, the underwater pressure exerted on an elephant seal at 600 m is equivalent to 60 atmospheres, that is, 60 times the pressure at the surface. At 1,000 m it is equivalent to 100 times the pressure at the surface. (We used to relieve boredom on long research ship expeditions by lowering Styrofoam head models to depth and then marveling at the strange, compressed, and contorted shapes created by the pressure when we reeled them up.) Diving elephant seals can ascend quickly from a deep dive, in large part, because they are not breathing air and nitrogen like Scuba divers. Nitrogen does not accumulate in the blood and bubble out to collect in the joints, blocking circulation and causing pain and paralysis. This enables the seal to avoid nitrogen narcosis and decompression sickness (the bends).

We know little about how elephant seals avoid the high-pressure nervous syndrome that humans get when diving to great depths. The symptoms include tremors, myo-clonic jerking, somnolence, EEG changes, visual disturbance, nausea, dizziness, and decreased mental performance. Studying the effects of pressure on seals in the conventional manner using a hyperbaric chamber is difficult because the chamber would have to be huge to accommodate an elephant seal, it would be expensive to construct, and it would be as dangerous as working with a bomb.

Anatomy

Over millions of years, operating in the marine environment has changed and molded the anatomy, physiology, and behavior of elephant seals, allowing them to better exploit the resources of the sea. Many of the changes improved hydrodynamics and made them more energetically efficient. After all, seawater is a dense medium that is 800 times denser than air and much more difficult to move through. The morpho-logical changes that occurred over millennia resulted in more streamlined bodies that moved through water with minimal expenditure of energy and reduced "drag" or resistance.

Limbs and feet were modified to fore flippers and hind flippers. Sex organs and mammary glands were retracted inside the body with only slits or nipples exposed to the outside (Figure 5.2). External ears were lost and replaced by small holes on each side of the head, eyelashes vanished and so did eyelids, replaced by eye mucus. Fur was reduced in length and could not be erected, as is so evident in domestic cats. Thick blubber around the body core provided a smooth, streamlined body shape and protection for internal organs. The large size and the blubber layer also provided insulation from cold ocean waters that ranged from 11.9° C at the surface to 4° C–6° C at depths of 400–1,000 m where females forage (Figure 5.3). Since water absorbs heat from the body 25 times more rapidly than air, maintaining core temperature of approximately 38° C was necessary.

To better modulate core body temperature, the seals developed an internal thermo-stat called counter-current heat exchange blood vessel system (Schmidt-Nielsen 1981). This is simply a network of arteries and veins in the flippers and skin or the nasal turbinates where there is no blubber (Huntley et al. 1984). The vessels are

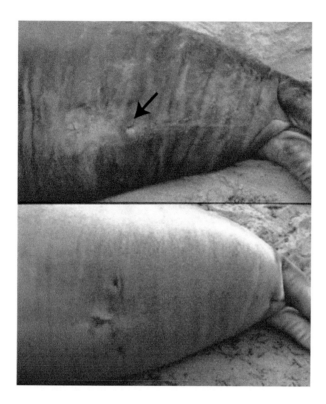

Figure 5.2 Recently weaned male and female pups. The male (top) is distinguished from the female (bottom) by a penile opening (black arrow).

Figure 5.3 A cross-sectional view of a dead weanling showing the blubber layer in relation to the rest of the tissue. Blubber makes up approximately 50% of the mass of seals at this age.

organized such that cool venous blood flowing to the heart is warmed by arterial blood going to external areas of the animal allowing the animal to retain heat.

Another problem for deep diving elephant seals is the virtual absence of light at the depths to which they dive routinely. In fact, they cope with this quite well because they have excellent vision (Levenson and Schusterman 1997, 1999; Carlson and Le Boeuf 1998). They see well at low light levels because they have more rods than cones, and they possess a structure that reflects light back to the retina, making it easier for the photoreceptors to detect light. In addition, females feed primarily on bioluminescent prey; the light generated by their prey signals their location (Campagna et al. 2001). Moreover, cold waters of 4° C–6° C make their prey sluggish and easier to capture and consume (Hakoyama et al. 1994).

They do not swim at the surface. They "yo-yo" dive down and up from point A to point B (Le Boeuf et al. 1992; Crocker et al. 1994). The descent portion of dives are relatively shallow, 30° to the bottom of the dive at a depth 400–500 m, followed by a steeper ascent of 54° to the surface where the animal breathes for two to three minutes then repeats this pattern over and over again until it arrives at the feeding location or runs into a prey patch (see Chapter 9 for details). The distance covered on a single dive is about 1.2 km. Adult males may continue this pattern over and over again on long migrations that last up to 40 days. Tuna and some other fishes do this as well for direct travel except that they do not surface (Carey and Scharold 1990; Holland et al. 1992; Block et al. 2001).

Why such a bizarre way of traveling? Yo-yo swimming is an energy-efficient mode of travel (Weihs 1973; Williams et al. 2000; Davis et al. 2001; Hassrick et al. 2007; but see Davis and Weihs 2007). If the animal is slightly negatively buoyant, it exhibits burst and glide swimming; it coasts down aided by gravity, covering 50% or more of the horizontal distance with reduced effort. The ascent portion of the dive requires more effort, but the distance the seal swims actively is short, less than if it covered the same horizontal distance entirely at or near the surface. Moreover, since the metabolic rate decreases during dives, swimming this way uses up less energy than swimming at the surface. To put this in perspective, swimming is a more efficient way of moving about than flying or running (Schmidt-Nielsen 1972). This is ironic because water is dense, but the difference is that a body in water requires little or no support and all effort can be directed to moving forward.

Behavior

Physiology and anatomy set limits on the diving behavior required for efficient exploitation of the sea for food but, ultimately, the behavior of getting food is the important end result. The diving behavior of elephant seals is extraordinary and pushes performance limits to extremes. In brief, the seals dive deep, long, and continuously while at sea. Diving behavior, migration, foraging, and diet are discussed in detail in Chapter 9.

In summary, over time elephant seals have become veritable diving machines, custom designed to operate efficiently and to acquire ample food in the marine environment. They have a virtual monopoly on the food source where they feed because they have few competitors; the niche is wide open because few animals can go where they go or do what they do.

Life on Land

In the process of adapting to forage at sea, however, elephant seals lost attributes that were useful and necessary on land. Most obvious is that they lost the ability to move quickly, which made them vulnerable to carnivorous land predators such as bears, wolves, and coyotes. Flippers serve swimming but are poor substitutes for feet and legs to run from danger. Elephant seals are slow and ponderous on land. Humans can outrun them, so long as they don't trip or fall. The pups and weanlings are the slowest, and given their body composition, about 50% lean tissue and 50% fat (Figure 5.3), they were a tasty and nutritious morsel for grizzly bears, *Ursus horribillis*, that roamed the coast of California searching for meat of any kind before they were wiped out by hunters in the 1870s. In the early part of the nineteenth century, the bear population was decimated at about the same time as the elephant seal, fur seal, and sea otter populations were being hunted into near oblivion. The wolf population went the same way.

The point is that giving birth on the mainland before 1850 would have been unwise for a mother seal. Grizzlies were a scourge and hazard to everyone, humans included. There are reports of as many as a dozen grizzlies feeding on or in the dead carcass of a beached gray whale on the California coast. We presume that this threat pressured elephant seals to breed on islands to avoid carnivorous predators. Islands may have provided some protection from the human aborigines as well, for whom they would have been easy prey.

In 1975, however, the recovering population of elephant seals started to breed on the Año Nuevo mainland (Le Boeuf and Panken 1977). Breeding followed on mainland sites at Point Reyes in 1981 and at Piedras Blancas near San Simeon in 1992 (Allen et al. 1989; Lowry et al. 2014). Evidently, a few pioneering seals tested these locations and deemed it safe, perhaps sensing that grizzly bears and wolves were no longer present in California. Not only was the predominant danger no longer present, on the contrary, there were rangers in their distinctive hats and green military-like garb protecting them from all other kinds of disturbance or harassment, especially from humans. Humans and elephant seals do not mix well on the same beach; they pose dangers to each other. Before rangers exerted control at Año Nuevo State Park, tourists would throw rocks at sleeping seals to get action photographs. I saw one tourist try to place his juvenile son on an elephant seal's back so he could get a picture! Rangers or docents were necessary to protect the seals from human harassment and the humans from being injured. In any case, the seals adapted and found new, ample, and relatively safe breeding sites. Breeding on mainland sites worked out well for them because their island rookeries were getting crowded and there were few other islands that were suitable.

So, all seemed well and colonies on mainland sites were increasing rapidly. On February 26, 2020, however, Sarah Codde (pers. comm.) reported that 26 of 50 weaned pups on one of the beaches at the Point Reyes rookery had fresh bites on their hind flippers or huge chunks of their hind flippers chewed off. The wounds varied from small chunks of flesh removed to approximately half of the hind flippers removed (Figure 5.4). Otherwise, the pups appeared healthy and fat as they should at this time of year. The pups were newly weaned and had not yet entered the water to begin learning to swim and dive. The bites were caused by coyotes. A pup at this stage of development can do little to prevent a coyote from taking a bite or chewing off flesh from its hind flippers; it cannot move away fast enough. Moreover, a weanling with half of its hind flippers removed, their main source of propulsion, would have a difficult time diving for food. Evidently, predators on the mainland remain a problem for breeding elephant seals.

Another problem for the elephant seal on land is that the blubber layer that is so necessary at sea can be a burden on land, making it easy for the seals to get overheated.

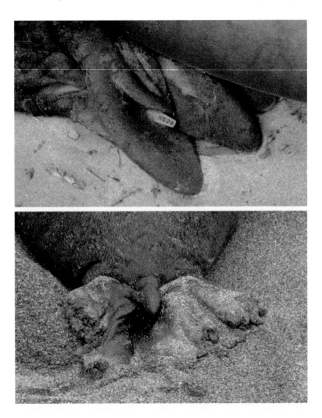

Figure 5.4 The hind flippers of a newly weaned pup showing that it was born at the Año Nuevo colony (top). The hind flippers of a newly weaned elephant seal (bottom) that have been bitten and chewed off by coyotes at the Point Reyes colony. Courtesy of Sarah Codde; Photograph by Marjorie Cox. (A black and white version of this figure will appear in some formats. For the color version, please refer to the plate section.)

Figure 5.5 Sand-flipping helps seals expose cooler sand below the surface and control their temperature on a hot day at Pilot Rock Beach on Isla de Guadalupe. Sand-flipping is also indicative of distress or nervousness. (A black and white version of this figure will appear in some formats. For the color version, please refer to the plate section.)

Blubber is about half the weight of a newly weaned pup (Figure 5.3). The blubber layer in adult females immediately after giving birth is as high as 40% of their mass (Crocker et al. 2001). Given that the mass loss of males over the breeding season is similar to that of females, the percentage of blubber of adult males at the beginning of the breeding season is presumed to be a similar proportion of total mass as that for females. But it takes a great deal of energy for males to move their massive bodies, and it is especially costly to do so over uneven terrain (Crocker et al. 2012). So, it is not surprising that the seals are least active when it is warm and sunny, even during the middle of winter. They have other ways to regulate temperature (White and Odell 1971; Heath and Schusterman 1975). They will flip sand in the air and over their backs with their fore flippers (Figure 5.5). This has a cooling effect because the seal is digging into wet, cool sand and the sand coats their backs exposed to the sun. The counter-current heat exchange blood vessel system used at sea to keep warm is also used on land, in reverse, to dissipate heat and maintain core body temperature (Schmidt-Nielsen 1981).

The positive side of wearing the heavy coat of blubber is that it provides the energy for everything they do on land: locomote, fight, nurse, and mate. While the most important thing at sea is feeding and fattening up the blubber layer, this fat coat is a boon and a burden on land. On the positive side, the blubber is like a well-stocked refrigerator full of food that each seal carries and feeds off of as necessary. All of the seals fast from food and water for the entire time spent on land and they can do this because they are "feeding" off of their blubber reservoir (Box 5.1). Consider this: the length of stay on land for dominant males fasting from food and water throughout the breeding season is up to 100 days. Clearly, conservation of this energy reservoir is critical. Saving energy comes in many forms. Here are two more of them. Large turbinates in the nose act like blotters squeezing out the moisture in each exhalation to

> **Box 5.1** The Jellybean Experiment
>
> One day in 1970, I was walking from the blind that overlooks the large elephant
> seal harem on Beach 17 at Año Nuevo Island on my way to cookhouse. Along the
> way, I passed a non-competitive subadult male dozing all alone in the dunes. As
> I approached, he emitted a relaxed, long drawn out yawn with his mouth fully
> agape. For no reason except that it was a target, I flipped a jellybean into its mouth.
> Two weeks later, I passed by the same spot and the subadult male was still there.
> He opened his mouth in a low-level threat as I approached, and there was the
> jellybean on his tongue looking the same as when I first deposited it there.
> Conclusion: the seal had not eaten, salivated, or swallowed in two weeks. It
> follows that seals, which don't eat during their long stay on land during the
> breeding season, do not defecate during this time. This makes them different from,
> and more pleasant to study than, sea lions that eat daily.

keep water loss to a minimum, because the only way to get water is to metabolize fat
(Huntley et al. 1984). The other one involves sleep on land, which mimics diving
behavior to such an extent that we called sleep on land "terrestrial dives." The seal holds
its breath for 15–20 minutes or more, depending on its age, then breathes for 2–3
minutes, and then repeats the pattern over and over again unless it is disturbed or has
slept enough (Blackwell and Le Boeuf 1993). The seals sleep while holding their breath
and are awake when breathing. You may see a seal sleeping with its head completely
submerged in a tidepool. It is not dead. It will lift its head after a while to take a few
breaths, yawn, and then put its head back down in the water and sleep for another
20 minutes or so. When the seal holds its breath, and its face is submerged in water, its
heart rate slows, and you can count the beats on the animal's chest; it is especially
obvious with the larger males. Moreover, irregular breathing with long apneas means
fewer exhalations than would be the case if the animal breathed regularly. This means
less water loss and results in a net energy savings of approximately 23%, varying with
the amount of time in the day the animal spends in apnea. Finally, consider how this
evolved. Elephant seals took extended breath-holds or apnea, originally selected for
pups to survive the birth process, retained and expanded this behavior in adulthood as a
fundamental adaptation for procuring prey while diving and also for conserving water
and energy during prolonged fasts on land. Remarkable, isn't it?

In summary, living in two different worlds is a balancing act for elephant seals;
they are exceptionally adaptable in this regard. Nevertheless, adapting too much in
one world compromises them in the other world. One presumes that this balancing act
is constantly being adjusted. That is, adaptations to the two different worlds are tightly
connected. Elephant seals manage well in both worlds at this time, but this may take
considerable juggling as the climate changes and the globe warms. Obviously,
elephant seals do extreme things in both habitats. These things are not qualitatively
different from what other air breathing and warm-blooded mammals do, but they
push things to extreme limits. They are quantitatively different. They do more, go
deeper, stay longer, and withstand more pressure. They do superlative things.

6 The Cost of Living in a Seal Harem

Everything costs something.

Zara Hairston

Life is tough in a crowded seal harem, especially for pups (Le Boeuf et al. 1972, 1994; Le Boeuf and Briggs 1977; Reiter et al. 1978, 1981; Riedman and Le Boeuf 1982; Le Boeuf and Condit 1983; Le Boeuf and Mesnick 1991; Reiter and Le Boeuf 1991; Rose et al. 1991; Le Boeuf and Campagna 1993). Group living gives rise to problems for those motivated by self-interest. Nevertheless, a female elephant seal is more likely to produce a viable pup surrounded by other females in the harem than by giving birth alone. We would not have animals living in colonies unless the advantages out-weighed the disadvantages.

Nevertheless, a female's neighbors in the harem are not her friends; they are her competitors, each of whom is most concerned with the health of her own pup. Newborns are the most vulnerable. Pups emerge wild-eyed and alert into a frenzied, noisy, chaotic maternity ward. The newborn wobbles and struggles to keep its head upright while its mother emits urgent pup attraction calls in its face to establish the mother–pup bond. Western gulls shriek and squabble overhead, fighting to feast on the placenta as soon as it is expelled. Neighboring females, protective of their own personal space, threaten the mother and may charge and bite her or the newborn. Other neighbors are aroused and call their pups urgently or approach the newborn to make sure it isn't theirs. Adult males careen through the harem, like freight trains, trampling everything in their way, and killing some newborns, while running roughshod over an intruder or fleeing with urgent abandon from their superiors. Storms with strong winds whip up high surf that inundates the rookery, causing the seals to bunch up and increasing female aggression. In this pandemonium, pups are often injured, orphaned, or washed out to sea. The transition from the womb to the outside world is a rude awakening.

Social living among elephant seals has its own advantages and disadvantages. Females that group get relief from suitors that pester them endlessly if they are alone. When females group, a male has the opportunity to mate with multiple females. Life in a harem is toughest for pups and their mothers. If a female's pup dies, this affects a large component of her reproductive success. The entire year is lost and subtracts from her lifetime reproductive success. Infant mortality is the most obvious cost of living in a seal harem. As the size of the social group increases, so do conflicts of interest between individuals and possibly parasites and disease transmission (Alexander 1974; Hoogland and Sherman 1976).

The annual pup mortality rate on the rookery before weaning varies with location, time in the breeding season, density of animals, storms with attendant high surf and tides, and space for the animals to move from high waters to higher ground (Le Boeuf and Briggs 1977). On Año Nuevo Island, the annual pup mortality rate in the years 1968–1976 varied from 13% to 26% of the pups born. This changes drastically in a bad weather year, however. For example, in the El Niño in 1983, the storm surge, high surf, and high tides all came together in perfect unison at peak season (Le Boeuf and Reiter 1991). The high surf flooded seal nurseries on the island and swept newborn pups out to sea where they drowned, or waves hurled them onto the rocks or simply separated them from their mothers, their source of nourishment (Figure 6.1). It was chaos. Seventy percent of the 975 pups born died (Le Boeuf and Condit 1983). You might think that seals, adapted to life in the ocean, would be well suited to survive being tossed into a raging, turbulent sea. On the contrary, thousands of pups from the several rookeries along the west coast of California and Mexico died in 1983; it was the greatest tragedy to befall these animals since their near extermination by commercial hunters in the last century. Obviously, storms at peak season change everything and the pup mortality rate soars.

Nevertheless, the pup mortality rate on the mainland, even in bad storm years like 1983, is far lower than on the island, ranging from about 8% to 11% of pups. The lower incidence on the mainland is because the seals could move away from the water to higher ground; there is no higher ground to move to on the island.

Dead pups are seen on the breeding beaches as early as mid-December and as late as mid-March. The pup mortality rate increases with the number of females in harems. Consequently, peak pup mortality is associated with peak season, mid-January to mid-February, when the number of females in attendance is maximum. The mean age of dead pups is 12 days, less than two weeks old. More male than female pups die because more are born.

Most pup mortalities on the rookery result from a complex set of interacting factors that includes separation from the mother, injury from males or females, and starvation followed by secondary complications such as septicemia and pneumonia. This set of interacting factors is called the Trauma-Starvation syndrome. Mother–pup separation occurs under a variety of social circumstances: (1) the mother is displaced from her pup by an aggressive female shortly after giving birth, which interferes with the imprinting process and results in the mother not recognizing her pup; (2) the mother loses sight of her pup soon after birth; (3) the pup is injured shortly after birth, and it cannot respond appropriately and the mother abandons it; (4) the mother simply moves away from her pup at birth to inspect another pup in front of her, to threaten a nearby female, or for no apparent reason; (5) the pup strays from its mother while she sleeps.

A pup separated from its mother has no source of food, so it moves about the harem calling (Figure 6.2a). These distress vocalizations normally stimulate the

Figure 6.1 (a) Beach 17 on Año Nuevo Island at the peak of the breeding season in late January of 1983. (b) The same breeding beach on January 27, 1983, as extreme high tides and powerful waves flooded the area, causing significant pup mortality.

mother to roll over on her side in the nursing position. If a reunion does not occur in two days, it is unlikely to ever occur. The orphaned pup's calls become more urgent. It moves from one female to the next and starvation compels it to attempt to suckle any available nipple. In doing so, the orphan is exposed to the constant danger of being bitten by nursing females or squashed by bulls that charge recklessly after each other in the harem. It is important to state that males squash pups inadvertently because they are in the way and they have their own agenda; males do not bite pups. Nevertheless, orphans are driven out to the periphery, as if by centrifugal force, where they wander about emitting distress calls. Hunger compels them to keep re-entering the harem to attempt to steal milk from any nursing female, but the majority are chased out and bitten by mothers guarding their milk jealously for their own offspring.

When a nursing female hears a pup calling, she is likely to investigate it even though her own pup is next to her. After sniffing the alien pup the cow may simply retreat to her own pup and pay no further attention to it or she may threaten or bite it viciously on the nose and head. Some females may even toss the alien pup into the air a couple of meters away. Females strike fast and they may strike repeatedly. As the injured orphan turns to flee it may be pursued and bitten on the rump or hind flippers. Or worse, its squawking or fleeing movements may incite several females nearby to

Figure 6.2 Nursing females. (a) A nursing mother, separated from her pup by two females, calls to it. (b) This female allows four pups to suckle, one of which may be her pup. (A black and white version of this figure will appear in some formats. For the color version, please refer to the plate section.)

join in and pummel it. As many as five or six females may bite the orphan. I've seen episodes where several females pursued and bit an orphan repeatedly and the pup died soon after the mobbing.

Most nursing females treat orphans like parasites. Females cannot feed two pups and nourish them adequately, so if a female attempts to do this, one or both pups may die. If one pup survives to weaning, its weight is considerably lower than that of other weaners and its chances of surviving at sea are decreased. Therefore, females are ever alert to orphans trying to steal their milk.

The plight of lost pups has important consequences. Although the majority of females in attendance during the breeding season nurse their own pups exclusively, it is common for many females to lose their pups. This leads to a situation where there are many orphans and pupless females. Females adopt pups (Le Boeuf 1972; Le Boeuf and Panken 1977; Reiter et al. 1978, 1981; Riedman and Le Boeuf 1982), and adoption takes many forms.

Adoption, or parental care of any kind directed to alien or unrelated young, is important because it appears to be altruistic; the foster parent or alloparent assists others at its own expense (Hamilton 1964). Adoption and alloparenting are costly behaviors that are inconsistent with evolutionary theory (Wilson 1975). Although it is not clear why an animal invests in another's offspring, fostering behavior is widespread in nature and multiple explanations may apply (Riedman 1982).

In crowded harems, such as the one on Año Nuevo Island from 1977 to 1980, 24%–57% of pups born each year were separated from their mothers (Riedman and Le Boeuf 1982). Most mother–pup separations (44%) were caused directly by males moving through the harem or attempting to mate with nursing females. This disturbance caused the least prolonged separations. Twenty seven percent of mother–pup separations were due to pups wandering from mothers, often in search of an additional milk source, which they often secured. Thirteen percent of the separations were due to aggression from female neighbors; these caused the most protracted separations, especially if the mother was involved in the altercation. Ten percent of the separations were due to inclement weather such as high surf conditions at high tide; this estimate is probably low because it is difficult to monitor separations when so much is going on. Most separations and adoptions occurred when pups were less than two weeks old.

Some mother–pup separations occurred immediately after parturition. The mother became confused and moved away from her newborn, or neighboring females attacked the mother. or the mother attacked neighbors leaving her pup behind.

Separations in which the mother and pup were reunited lasted from less than one minute to as long as four days. Seventy seven percent of mother–pup separations lasted 10 minutes or less. Separation distance varied from 2 m to 40 m; most separations were of four meters or less.

Mothers attempted to reunite with their pups by emitting a distinctive pup-attraction call, a warbling, yodel-like vocalization in which the head is moved rapidly up and down (Figure 6.2a). Pups emitted the distress call to which the mother responded. Sixty-six percent of the time, the mother found the pup; 14% of the time the pup found its mother. Frequently, the mother enticed her wandering pup to remain near by rolling over into the nursing position; this often occurred when storms caused widespread disturbance in the harem.

The odds of an orphan surviving are low. In four years of monitoring 572 orphans, 1976–1979, the proportion that died each year before six weeks of age was 54%, 48%, 54%, and 66%. Five percent of orphans reunited with their mother, 27% were adopted or frequently cared for by foster mothers, and 68% were not adopted, rarely fostered, and most of these died.

The survival of an orphan depends on the amount of milk and care it receives from a foster mother. A variety of fostering behaviors are observed. The most common form is when a female that has lost her own pup fosters a single orphan. Some females adopt a pup while it is still with its mother; they simply settle near the mother–pup pair and act as allomothers. Some females aggressively attempt to steal a pup from its mother. Some allomothers remain with a pup several days after its mother weans it. A female without her own pup may adopt a pup after it is weaned by its mother. Some orphans acquire two foster mothers that take turns feeding it. Some foster mothers feed an orphan in addition to their own pup. Some females adopt two pups simultaneously or consecutively. Every year a few females allow five to six pups to suckle (Figure 6.2b). Some of these females still have their own pup, but it wanders away because it receives inadequate care and the mother does not distinguish it from other orphans.

Orphans bear the scars of being bitten while forced to steal milk. The few that survive to weaning age are underweight, and their prospects for survival at sea are low. Those that have the good fortune of being adopted by females that lost their own pups are fostered like filial pups until weaning. From the female perspective, infant mortality is a cost that diminishes a mother's reproductive success.

Clearly, adoption is a common behavior in elephant seals, particularly among females that have lost their own pups. Moreover, an adopted pup's age is highly and positively correlated with the age of its foster mother's filial pup. This makes sense because the fat content of a female's milk increases from 15% at parturition to 55% during the final week of nursing (Riedman and Ortiz 1979), and consequently the foster mother's stage of lactation must be in sync with what the pup can incorporate.

What is the benefit of adopting an alien pup? The allomother may be genetically related to the adopted pup, which would increase her inclusive fitness; the next best thing to having your own pup is adopting a relative with whom you share genes. We don't know that this is the case with elephant seals. Given the crowd of females breeding close to each other, and no evidence that they recognize kin, we think adopting close relatives is unlikely. Otherwise, pupless females gain valuable maternal experience in raising an adopted pup that can serve them well in later life. Maternal care involves considerable skill, especially in large, crowded rookeries. As stated in Chapter 8 on female reproduction, the weaning rate of female elephant seals increases dramatically over the first three or four times they give birth. Females learn how to be good mothers. Lastly, females that have lost their own pups are primed and ready to nurse and perhaps any pup of the right age will do.

There are still other dangers for weanlings and mothers in and out of harems. Most weanlings move out of the harems into the sand dunes nearby where they gather in pods numbering up to 20 or 30 individuals bunched together, sleeping by day and cavorting in the shallow waters at night. Since most females have departed the harems to return to sea and feed, libidinous subadult males treat the weanlings as female substitutes (Rose et al. 1991). They accost the weaners and try to mate with them. Many healthy weanlings are injured and killed by head and neck bites inflicted during attempted copulation (Figure 6.3).

Females leaving the harem have a similar problem. They must avoid moving through the gauntlet of horny males to reach the sea (Le Boeuf and Mesnick 1991). In the process some females are injured and killed. This behavior is described in Chapter 8, on female mating and reproductive behavior.

Males, too, are not exempt from paying a cost for living and competing in a seal harem. Harems are the battleground where they fight and threaten each other for access to females and bloody injuries are common (Clinton and Le Boeuf 1993; Deutsch et al. 1994). One breeding season, all males in residence were surveyed for

Figure 6.3 Subadult males injure weanlings. (a) A subadult male attempts to copulate with a weanling. Because weanlings are smaller than adult females, the stereotypical neck bite becomes a head bite. (b) Many weanlings have neck scars, a sign that they have been mounted by males.

bleeding cuts, bruises, and injuries from neck bites. Ninety five percent of the males had wounds caused by other males that ranged from moderate nicks and punctures to deep exposed wounds into the blubber layer. Surprisingly, infections to even serious wounds were rare. Female injuries from intersexual competition were significantly less frequent and less serious.

In summary, the highest cost of living in a seal harem is to pups that die primarily from starvation, infections, pneumonia, bites from females, and trampling by adult males. These proximal causes start with the pup being separated from its mother. Young pups less than two weeks old are the most frequent victims. The reproductive success of mothers, especially young ones, decreases if she loses her pup. Males incur numerous injuries from other males in competing in and around harems.

7 *Coito Ergo Sum*
Males Explained

Chi mi frena in tal momento?

Donizetti from Lucia de Lammermoor

If you visit an elephant seal rookery, like Año Nuevo or Piedras Blancas, during the breeding season, and you want to understand what is going on, keep your mind in the gutter. It's all about sex or its consequences, reproduction. The simple agenda for males is to mate, nothing more, stop, period, and to do it over and over again with as many females as possible (Le Boeuf and Peterson 1969b; Le Boeuf 1971a, 1974, 1984, 1991; Le Boeuf and Reiter 1988). In contrast, the behavior of females is geared to giving birth, weaning pups, and getting re-impregnated to give birth again next year; for females, the mating part is a brief encounter with a stranger she is not likely to ever see again.

What is the basis of the "social organization" that plays out on the beach? It starts with females grouping together during the breeding season, because if they did not, isolated females would be pestered incessantly by amorous males and giving birth to a viable pup would be virtually impossible. When females group together, however, the probability of any individual being disturbed decreases. Female clumping affects males; it gives a male the opportunity to mate with multiple females, as opposed to when females are widely distributed in the environment and it is a one-to-one proposition. When males compete to mate with as many females as possible, the strongest males prevail and monopolize mating. The discrepancy in mating success between the haves and have-nots is huge. Large size in males is selected because big males dominate smaller males, as a rule, and do most of the mating. Over time, males get larger and larger compared to females and we end up with striking sexual dimorphism, with adult male elephant seals being 3–10 times larger than adult females. The great discrepancy in mating among males results in extreme polygyny, with a few males mating with many females and most males never mating at all. They are highly motivated to mate but they cannot because they are prevented from doing so by males dominant to them, in other words, because of the power structure. In polygynous societies, the variance in reproductive success is greater among males than females.

Males are ready, willing, and able to mate when they go through puberty at five or six years of age, but the power structure – older, larger, and more powerful males – prevents most of them from mating until they are about eight. Nevertheless, they are a nuisance to nursing females and a burden for established males in the harem because they are relentless in attempting to sneak in and steal copulations. They are rarely successful.

Prime breeding age is 9–13 years (Clinton and Le Boeuf 1993). The peak for successful mating is age 12 followed by ages 13, 11, 10, and 9. Males 14 years of age are old and "over the hill"; they watch from the periphery of harems and don't even try to approach females. They do not survive beyond 14 years of age (Le Boeuf and Reiter 1988). Male survivorship to age 12 and 13 is approximately 1% of males born and 2% of males weaned.

Males are typically classified as follows: a 4-year-old male is a subadult male 1 or SAM1, a 5-year-old is a SAM2, a 6-year-old is a SAM3, a 7-year-old is a SAM4, and males 8–14 years old are adults (Cox and Le Boeuf 1977). Clearly, an important step in achieving high reproductive success for males is living to prime breeding years.

Males reach physical maturity at 8–9 years of age but may continue to grow until they are about 10 years old. A mature bull may weigh over two tons, that is, over 1,800 kg (Deutsch et al. 1990, 1994), and have a total length, from tip of nose to tip of hind flippers, of almost 5 m. Six-year-old males (SAM3 class) are roughly two-thirds the mass of mature bulls (Box 7.1).

Once males begin breeding, they have a relatively brief reproductive life span, two to four years at most. The reproductive life of females, on the other hand, may last up to 20 years. This is the big picture of the mating game in northern elephant seals, and as expected, the mating strategies of the sexes are dissimilar.

The breeding season begins with the arrival of adult males in early December; most of them will remain on the rookery until mid-March (Le Boeuf and Laws 1994b). Males come to the traditional beaches where females will arrive later to give birth, nurse their pups, and mate. Arriving males are in a bellicose mood, threatening each other and fighting to establish rank in a dominance hierarchy in and around a group of females, also known as a harem. Males compete to prevent access to females from the time they arrive to when they come into estrus about a month later (Figure 7.1). Dominance is expressed by threats in which the head and forequarters of the male are elevated and a series of loud, low-frequency, guttural, clap-like vocal bursts are emitted (Figure 1.1; Box 7.2). If the threatened male is subordinate to the aggressor, he moves away. If he does not move away or moves away too slowly, he is attacked and bitten savagely. Males have formidable canine teeth that extend about 5 cm from the gum line. If the threatened male responds with a threat vocalization of his own, a fight ensues in which each male elevates his head and neck and delivers blows to the neck of the other and attempts to bite the other on the neck. The biter grasps flesh and pulls away, leaving open wounds. If one of the fighters gets a hold on the other's proboscis, he holds on, worries it, shaking it from side to side, shreds it through his teeth, and usually wins the fight. The fights are bloody affairs, with most blows and bites directed to the neck shield. Fights may be as brief as a few seconds before one gives up and retreats, or as long as 25 minutes. The longest fights start on land and move into the water. The loser flees from the winner. The winner may give chase and bite the retreating loser on the back. When the pair next meet, the winner need only threaten the loser to displace him. Dominance has been established.

Box 7.1 Aberrations

If you observe thousands of seals over the years, or any animal, for that matter, you are likely to see some unusual physical and behavioral aberrations.

There was a seven- or eight-year-old male who threatened and fought with other males and vigorously mounted and positioned himself so as to copulate with females. Try as he might, he never achieved intromission. The reason was clear; he had no penile opening, which is positioned 6–12 inches south of the belly button. One could see that his penis was erect inside his body, but it had no way out. He was motivated and eager to mate but completing the act was impossible. How did he urinate? Was it through the anogenital area, rather like females?

Two other males, named Hermes and Reverse, also lacked the usual penile opening but rather had a penis that pointed backward peeking through the anogenital area. They behaved like normal males and tried to copulate but positioning the body to achieve insertion was impossible even when the female was proactive and willing. Talk about frustrating.

Two nipples adorn the body of female elephant seals. That is all you see. The mammary glands are widely distributed in the body, giving the animal a stream-lined appearance that makes swimming most efficient. Two nipples are normal and common but over the years we have seen several females with three nipples. It was not clear that these supernumerary teats were functional. Supernumerary nipples or breasts occur in many mammals and are well documented in both males and females in our own species. Supernumerary nipples are diagnosed in humans at a rate of approximately 1 in 18 people. So, this aberration is not especially unusual in seals.

Sexually mature male elephant seals are extremely vocal. Threats are issued when the male uses its forelimbs to elevate its head and neck and opens its mouth wide to vocalize. The vocal behavior of males can vary considerably from one individual to the next. We named one male with unusual vocal behavior after Spiro Agnew, the vice president of the United States from 1969 to 1973, a politician who was prone to making verbal gaffes when attacking his political enemies. Spiro, the seal, issued threat vocalizations with his mouth virtually closed. He "talked" out of the side of his mouth. This made it difficult to locate him and where the sound was coming from. The ventriloquial effect certainly caught our attention but it did not benefit the seal in competition with other males. While he did not resign from office in disgrace like his namesake, that is, quit competing, he never achieved a sufficiently high social rank to gain access to females.

The dominance hierarchy is virtually linear, A beats B, B beats C, C beats D, and so on, but there may be deviations from linearity, or triangular relationships, such as A beats B, B beats C, but D beats C, and both C and D beat E, and so on (Le Boeuf and Peterson 1969b). This happens because the relationships are between individuals. Evidently, males learn which male they can beat and which male beats them. They recognize each other by their threat vocalization or by sight. It is clear that they have

Figure 7.1 Two males fighting for dominance. (a–c) Each male attempts to tower over the other and to bite the neck or nose of the other. (d) The winner of the fight, shown here, is as bloody as the loser. (A black and white version of this figure will appear in some formats. For the color version, please refer to the plate section.)

the mental capacity to remember scores of competitors. Achieving high rank in the dominance hierarchy is vital because it gives access to females and mating, which is the highest priority of all males in attendance.

An important outcome of the dominance hierarchy is that if the alpha male loses a fight and is deposed, he does not simply go down a rank or two, he plummets down in the hierarchy to a level so low that he might not be competitive for the rest of the breeding season (Box 7.3). It is as if he was supremely confident as an alpha but once he loses a fight, he loses confidence. This is a common observation in other animals with dominance hierarchies, such as roosters in a barnyard (Guhl 1962). In human boxers, the champ is considered invincible, and the sparring partner is never allowed to hurt him or take him down.

The dominance hierarchy among resident males is established before the first females begin arriving in mid-December. The hierarchy, however, may change as additional males arrive and challenge the status quo.

Box 7.2 Training a Seal to Talk

In the early 1970s, an adult male was held in captivity at Marineland of the Pacific. It was one of several marine mammals living their lives in captivity. People paid to see marine mammals. The directors of the aquarium reasoned that it wasn't enough for customers to see them sleeping or swimming robotically in their pools; the paying public wanted to see them doing something. So, the dolphins were taught to jump through hoops that were placed higher and higher. A trainer would put his head into the open maw of a killer whale and the viewers would gasp. A sea lion balanced a rubber ball on its nose. Another seal would retrieve a basketball and put it into a hoop. All of this was reinforced with a food reward.

The bull elephant seal was taught to talk on command, that is, to emit the stereotyped threat vocalization it uses so frequently during the breeding season. This went well at first. The seal emitted a string of seven loud, guttural clap-threats and then waited for the reward. The number of clap-threats in a threat vocalization is relatively invariant among males in nature. Inevitably, they repeat the same string of clap-threats at the same speed each time. But as the training and the shows went on, our male at Marineland began to stint on the number of his vocal claps-threats before waiting for the fish award. At first, it was from seven to six and then from six to five and eventually to just one. He did only what was necessary to get the piece of fish. Smart? Effective? Whatever.

Observations in nature reveal that a male is smart about one other thing, remembering who he dominates and who dominates him. He does this by recognizing the other's threat vocalization, and possibly with the addition of visual cues. Consider that a male may have 50 or more competitors with whom he interacts daily around a large harem. He must instantly recognize each one of them and behave accordingly, chasing one away to confirm his dominance or avoiding the other to prevent being injured. Can you identify 50 of your friends or acquaintances immediately from their voices alone?

Arriving females are pregnant and associate with other females in dense aggregations or harems. The term "harem" refers to a social unit comprised of one or a few mature bulls, one of whom is dominant, the alpha, and a variable number of females numbering up to 200 or more. The alpha male does not keep a fixed group of females and he does not herd them, as some sea lions do. His task is to keep all males away from the females in the harem. He "wants" the sole privilege of mating with all of them. He fights for mating rights. The proximity of males to the female harem is determined by social rank; the alpha bull usually locates in the center of the harem or in the best position to prevent other males from approaching the females. As the females in the harem become more numerous and increase to a peak in late January, a point may be reached when other high-ranking bulls may also locate in the harem so long as they avoid the alpha male. Once there are over a hundred or so females in the harem, the alpha bull has difficulty keeping other males from entering. When this

> **Box 7.3** The Fall of a Titan
>
> He was the alpha bull at Año Nuevo Point, location of the oldest and largest harem in the entire area in 1985. He ruled with an iron fist. Just as sure as the sun came up in the morning, you know that he would dominate tomorrow. He was a prime example of the *und so weiter ad nauseam*. One knew what to expect. The power structure was stable . . . until the day he fought with a challenger. But this time, the challenger did not back down. They fought long and hard, and when it was over the alpha was bloody and immobile. The next morning he was lying dead on the very same spot. The king was dead; long live the new king.
>
> What happened over the next week reveals a lot about the inner life of these males competing with each other for the most important thing in their lives, access to females or mating rights. As males continued to move about the harem during the day, attempting to steal copulations, chasing others, and being chased by them, when a male passed by the moribund former alpha, he would stop, bite into the dead seal, and shake his head violently to punish the body of the seal that formerly abused him so much. This went on for a week, with numerous males showing the same behavior. Evidently, they remembered the former dictator well and it felt good to unload their animosity.
>
> It is hard to say how frequently males die from injuries sustained in fights. It is likely that a mortally wounded male may slink off and die far away, at sea, or many days later. It is extremely rare for a male to die from a fight on the spot.

happens, he occupies a central place that allows him to mate with as many females as possible. The beta and gamma males in the harem adopt a similar strategy while avoiding their superiors. The lower a male's rank, the more difficult it is to penetrate the defense of higher ranking males and to gain entrance into the harem. Each bull is aggressive in keeping males lower in rank than himself from approaching females.

Copulation is always initiated by the male (Le Boeuf 1972)(Figure 7.2). A male's mating attempt is not a gentle overture. He does not dance, sing, or attempt to attract or command the female's attention. The mating attempt is aggressive, direct, and reflects simply what the male wants. It can be potentially life-threatening to the female, because males are considerably larger than females and females may be vulnerable depending on their condition, for example, pregnant or underweight at the end of nursing. A male intent on mating does not investigate the head or anogenital region of the female as a possible sign of estrus, as males of many species do. Rather, he moves directly to a female, puts a fore flipper over her back, pulls her strongly toward him, and attempts to establish genital contact. If she resists, he pins her down with the weight of his head and neck and grabs hold of her with a neck bite to keep her in place. An unreceptive female struggles and resists by emitting a loud, croaking vocalization, flipping sand with her fore flippers back toward the face of the male, and swinging her hindquarters back and forth laterally, causing her hind flippers to smack hard against the male's ventrum and partially extruded penis. Males usually persist despite the protestations of the female, provided that they are not chased off by another male.

Figure 7.2 Copulation, when uninterrupted, lasts about five minutes. Note the size difference between the pair.

As for the mechanics of mating, copulation lasts about five minutes on land and a bit longer in the water. Most copulations occur on land. Some beachmasters have been observed copulating almost every day during the mating period. We've recorded a male copulating as many as nine times a day during daylight hours. Another male copulated 13 times on the periphery of the harem between 0922 and 1450 hours and all copulations were with the same female! Late in the breeding season it is common for males that mate often to show evidence of erectile disfunction. They mount females but cannot achieve tumescence. It is not clear if frequent mating causes a decline in sperm number, motility, or viability, as is the case with terrestrial mammals such as dogs.

Males are indiscriminate about mating partners. They will attempt to mate with pregnant females, non-estrous females that are nursing, females that have already mated multiple times with other males and are trying to depart the rookery to feed, dead females on the beach, and newly weaned pups,

Mating occurs at all times during the day and night. That is, males spend most of their active hours trying to copulate or preventing others from doing so. Interrupted copulations are frequent. The probability that a mounting male will be interrupted by another male is a function of the mounter's social rank (Cox and Le Boeuf 1977). The lower his rank, the higher the probability of interruption. The result is that mature males of high social rank have more time and freedom to attempt copulation, and this helps them to monopolize mating.

After mating, the male's relationship with his mate is over, unless he mates with her again and again, which occurs in certain situations. A male may sire many pups in his life, but he is a father in the genetic sense only in that he provides half of the genes of the offspring. He does nothing else to promote the survival of his progeny. Indeed, he does not know if a pup he may have sired last year survived, and if it did, he would not recognize it. In harems, males ignore pups in pursuing their agenda. They injure pups in their path by moving over them rather than avoiding them, or they may come to rest oblivious of the squawking pup trapped under them.

What is the upshot of this male strategy to mate at all costs? Research studies show unequivocally that social rank is positively associated with mating success. The alpha male associated with a harem monopolizes mating with the majority of females or all of them when possible. Others that succeed in mating are the highest-ranking males below alpha. Two studies support these conclusions. The first (Le Boeuf 1974) was conducted from 1968 to 1973 at Año Nuevo Island, when there were 193 to 470 females in harems. Less than a third of the males in residence copulated during each breeding season. A few males were responsible for the majority of the copulations observed. The percentage of males copulating increased with the number of females present, from 14% in 1968 to 34% in 1973. Similarly, the percentage of mating by the five most-active males decreased as the number of females increased, from 83% in 1968 to 48% in 1973.

Clearly, the number of males mating, or the mating success of a particular male, is limited by the number of estrous females present in the harem, the number of competing males, and the harem location and topographic setting. That is, a male's mating success varies with his ability to defend the harem. The larger the harem size, the more area it covers, and the more the male competitors, the more difficult it is for the dominant male to keep other males out. Exposed harems are harder to defend than those that afford restricted access to males.

As harem size increases, younger males are able to infiltrate harems and steal a few copulations. Those that succeed are sneaky and keep a low profile to avoid being discovered by their superiors. Mating success early in life, even though marginal, has an adverse effect on survival. These males are less likely to survive to the next breeding season.

In this particular study, mating success correlated positively with age and social rank. Copulation frequency was roughly proportional to social rank; the higher a male's social rank, the more frequently he mated. For example, the alpha male achieved the greatest percentage of the matings observed, which ranged from 12% to 37%, depending on the year of the study, harem size and location, and colony composition. In small harems of 25–50 females, the alpha might be the only male to mate.

One male dominated breeding for up to four consecutive seasons! Adrian was the alpha male during the four-year period 1971–1974. During this time, we estimate that he inseminated over 200 females. Some other top-ranked males succeeded in copulating in three to four breeding seasons. Most males, however, died within a year or two after their most successful year of mating.

The reproductive success of most males was nil or low because most died before reaching breeding age. For example, 75% of males born died by five years of age and by the age of nine, only 4% of pups weaned remained alive. It is also notable that surviving to breeding age alone was not sufficient to achieve mating success; several males were observed to mature, peak, decline, and die without ever mating. They tried to mate but were prevented from doing so by higher ranking males.

Individual strategy has an important influence on mating success. Males that spent much energy early in the breeding season were often exhausted and weakened by midseason and lost a critical fight and social rank. Arriving late in the season was also a detriment to mating because the social hierarchy was already in place, and these males had to fight too many battles to achieve a high rank. The optimal strategy for maximizing mating success was for a male to arrive early in December, obtain one of the top five social ranks, rest a lot, and minimize energy expenditure until mid-January when females began to come into estrus, and at this time move aggressively to achieve the alpha male position.

The results of a subsequent study were similar (Le Boeuf and Reiter 1988). Lifetime mating success of 138 male pups weaned in the years 1964–1967 was determined. All of them were observed until they died. Most males in the sample (77%) did not survive to puberty at five years of age. Only nine of the survivors, 6.5% of the sample, mated. They were estimated to have inseminated 348 females. The most successful male was estimated to have inseminated 121 females or 34% of the females; the next two males inseminated 97 and 63 females. The other males inseminated 3–63 females. The peak age for mating success was age 11, followed by ages 10 and 12. No male mated after age 12 although one survived to age 14. No male lived beyond age 14.

Both of these long-term studies show that male mating success is highly variable. A few males are highly successful. A few more are moderately successful. The majority fail to breed. Surviving to breeding age is critical. The chance of living to the age of high mating success is low. The peak age for mating success is 11–12; these males get the lion's share of copulations. After that comes males of age 13 and then those that are 9–10 years old. Mating success declines precipitously in younger males of ages 8, 7, 6, and 5. Males are too young to mate at ages 1–4 and too old to mate at age 14. Male survivorship to age 12 or 13 is approximately 1% of the males born.

What is the cost to males of the immense reproductive effort over the three-month breeding season, while fasting from food and water? Reproductive effort is defined as the proportion of available resources (time and energy) that an organism devotes to reproduction over a specified period of time (Gadgil and Bossert 1970). It is important because it measures resources allocated to reproduction at the expense of growth and maintenance (Box 7.4).

Several measures reflect the reproductive effort of males and the high price they pay to mate:

(1) *Energy expenditure.* Adult males expended energy during the breeding season at a rate of 195 ± 49 MJ/d, which is equivalent to 3.1 times the standard metabolic rate predicted by Kleiber's equation (Crocker et al. 2012). Kleiber's law states that for the vast majority of animals, an animal's metabolic rate scales to the ¾ power of the animal's mass (Kleiber 1932). Energy expenditure, depletion of blubber reserves, and water efflux were greater in alpha males than subordinate males. That is, energy expenditure increased with dominance rank and mating success. Daily energy expenditure was 267 MJ/d for highly ranked males compared to 155 MJ/d for peripheral males. High-ranking males experienced increased cutaneous water loss owing to frequent threat vocalizations. On average, males met 7% of

their energy expenditure through protein catabolism and 93% through fat catabolism. Terrestrial locomotion and the topography of the breeding site had a strong influence on energy expenditure. It takes a lot of energy for males to move their massive bodies. Total reproductive effort is significantly higher in males than females.

Box 7.4 Weighing Males

We wanted to estimate the energy spent by male elephant seals in their effort to reproduce, and weight loss provides a useful index. We weigh a female by drugging her, roll her into a plastic tarpaulin, and hoist her up to a scale attached to a tripod. But we couldn't use this method of weighing because the males are too big.

Two adventurous graduate students took up the challenge (Deutsch et al. 1990; Haley et al. 1991). Unrestrained bulls were weighed on a platform scale made of aircraft-strength aluminum that measured 4.9 m long, 1.2 m wide, and 0.13 m high. "Wings" measuring 1.2 m long had to be added to widen the scale to accommodate the seal's pectoral flippers. Two weigh bars containing load cells supported the platform, and cables ran from the weigh bars to an electronic display some distance away. The scale was placed in a site near harems bounded closely by dunes or willow thickets that acted as a natural channel. One way to lure bulls onto the scale was with a life-size, urethane model of a female elephant seal and the simultaneous playback of vocalizations of a female being mounted by a male. This caught the attention of males up to 50 m away and they approached and followed the model as it was pulled over the scale. The seal was stopped on the scale by raising a remotely operated snow fence in its face, sort of like what Rube Goldberg would do. The seal had to remain on the scale for 5–10 seconds to obtain an accurate reading. To make things even more difficult, one had to obtain at least two weight measurements for each seal to estimate total percentage of mass lost over the breeding season. Assuming a constant rate of mass loss throughout the time ashore, we estimated mass by extrapolation back to the arrival date and forward to the departure date. The average mass of adult males averaged 1,704 kg. The largest male in the sample, an alpha male for five years, weighed 2,265 kg, or about 2½ tons, when he first appeared on the rookery in late November, which is close to the weight of the entire offense and the defense of a professional football team, 22 players! This male lost 1,049 kg during his 106-day fast, or 46% of his arrival mass.

The mean percentage of arrival mass lost by all males over the breeding season was slightly over a third of their arrival mass and ranged from 26% to 46%. The most dominant males lost the most mass, which is understandable because they are the most active males on the rookery. Clearly, the effort devoted to reproduction by males is immense.

It is notable that adult males lose a similar percentage of their body mass, on average, as females (36%) over the course of the breeding season, despite the great sexual disparity in body size, activity, and duration of reproductive effort.

Box 7.4 (*cont.*)

Figure B7.4.1 Weighing males. (a) A subadult male is lured onto the scale by a decoy female emitting protest calls that she is being mounted. (b) An adult male is discouraged from moving further as he pursues the decoy and rests momentarily on the scale. (c) An adult male pauses on the scale. (d) The male's weight is recorded remotely as he rests on the scale. (A black and white version of this figure will appear in some formats. For the color version, please refer to the plate section.)

(2) *Mass lost during the breeding season.* Mass loss is a useful measure of reproductive energy expenditure (Fedak and Anderson 1982), especially in elephant seals because they fast and remain on land during the breeding season. See Box 7.4 for details on weighing males. Adult males weigh up to 2,265 kg when they arrive on the rookery at the start of the breeding season (Deutsch et al. 1990, 1994; Haley et al. 1991). In general, males lose an average of 36%–46% of their arrival mass while fasting over the three-month breeding season. Some bulls lose up to half of their arrival mass. Daily mass lost was positively related with body size, age class, dominance rank, mating success, and activity. Estimates of total mass loss over the entire breeding season varied from 198 kg for the smallest subadult to 1,049 kg for the largest adult. Put another way, mass lost per day ranges between 3 and 8 kg/day for subadult and adult males (Figure 7.3). That is, mass lost per day is greatest for the dominant males and decreases with dominance rank (Figure 7.4). Over the breeding season, mass loss ranges from 700–500 kg for young subadult males to 2,000–1,300 kg for the largest males (Figure 7.5). Mass lost during the breeding season is more than twice that of other

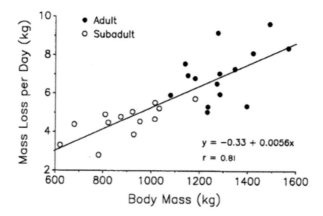

Figure 7.3 During the breeding season, mass lost per day by males increases with body mass, being higher in adults than subadult males.

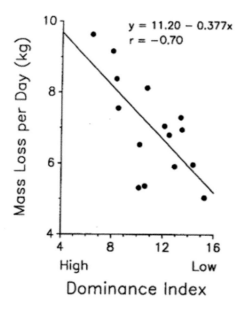

Figure 7.4 During the breeding season, mass lost per day by males increases with dominance rank.

pinnipeds. Overall, males and females lose approximately the same percentage of their arrival mass (36%), but the period is condensed to a month in females who are nursing and three months in males that are fighting and mating. Evidently, the sexes deplete their energy reserves to similar relative levels.

(3) *Injuries to males from fighting with conspecifics.* Injuries provide an index of risk taken to reproduce and show the degree to which males are exposed to potentially damaging or lethal interactions (Deutsch et al. 1994). In a single breeding season,

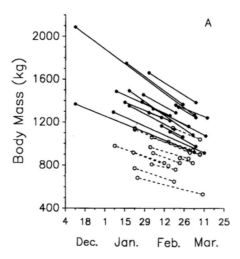

Figure 7.5 Body mass of adult (closed circles) and subadult males (open circles) as a function of date in the breeding season.

 90% of the competing adult males bore fresh wounds that were open, bleeding, or oozing from recent bites incurred during aggressive conflicts with competitors. Subadult males, seven to eight years of age, were especially vulnerable to injuries. In comparison, less than 20% of females had fresh wounds; there were fewer wounds on the body, and they were less severe. Fresh wounds on the head (including the face, nose, and eye) were significantly higher for males than females.

(4) *Sleep.* Sleep deprivation is associated with reproductive effort. Males forego sleep to keep others from breeding or to avoid being caught unawares and injured by other males. They must catnap when they can, usually this means when all other males are sleeping. At the end of the breeding season, when all females have departed the rookery, the males who have been fasting from food and water for about 100 days do not rush off to sea to feed. Rather, they sleep on the rookery for up to two weeks before going off to forage, which indicates that recovering from sleep deprivation is more important than recovering from the long fast.

(5) *Survival.* Early studies (Le Boeuf 1974) showed that males were more likely to die in the year following successful mating, suggesting that the reproductive effort was so great it lowered the probability of surviving. This was confirmed by Deutsch et al. (1990). Recent data with a much larger sample of males (Kienle 2019) show that death at sea following the breeding season is twice as high in males as females, indicating that male breeding effort comes at the cost of future survival.

The general conclusion is that male reproductive effort is costly for both species, and the most dominant males expend the greatest reproductive effort. Obviously, the time at sea following the breeding season is an important time to recover energy lost in their effort to reproduce.

In a different arena, the libido of males and its expression have dangerous consequences. Males are potentially dangerous to females, pups, and juveniles because of their great size, large canines, and habit of biting the neck when attempting to mate. Females departing the rookery at the end of lactation are especially vulnerable (Le Boeuf and Mesnick 1991). They leave the relative safety of the harem and the control of the alpha male to reach the water, forage, and recover from giving birth and nursing. To do so, they must run a gauntlet of peripheral males that hang out on the edge of the harem (Figure 7.6). No matter that these females have already mated in the harem and all of them are probably already inseminated. The waiting males are horny, having been denied access to females in the harem by the power structure. The departing female is alone and presents a mating opportunity for these males, usually from 6 to 12 of them. They are quick to compete with each other to stop the female's progress and mate with her. Bedlam ensues. Each pursuing male alternates between fighting off competitors and stopping the female's progress. The highest-ranking male has the advantage, but he must mate quickly before he is interrupted by an even higher ranking male. In this situation, the male pounds the female with the force of his heavy head and neck to stop her and then bites and holds on to the female's neck to control her escape. The female is especially vulnerable at this time because she has been fasting for over a month while nursing and has lost considerable weight (36%–42% of her mass). Her blubber layer, her coat of armor, is thinnest at this time of year, meaning less protection from blows to the body, especially for the large extradural vein that runs the length of the spine. Misdirected neck bites, when the female is moving to get away, can result in the male's large canines penetrating into the female's head, neck, or back. Bites to the head cause lethal brain damage. Bites to the back may puncture the large extradural, intravertebral vein. Head and body slams can cause broken ribs, organ damage, and internal hemorrhage (Figure 7.7).

Figure 7.6 Four peripheral males chase a female (black arrow) as she departs the harem. The oldest, largest male is in position to block her escape and mate with her. (A black and white version of this figure will appear in some formats. For the color version, please refer to the plate section.)

Figure 7.7 A female with a bloody head wound caused by a peripheral male attempting to mate with her as she left the harem to go to sea. (A black and white version of this figure will appear in some formats. For the color version, please refer to the plate section.)

We determined that 17 females were inadvertently killed by courting males during a 20-year period as they left the harem for the sea. The incidence of female mortality due to this male behavior was about one death per thousand females giving birth. Although the incidence is low, its importance is not diminished. The probability that an individual female is victimized is much higher when you consider that she may breed up to 20 times in her life. Moreover, the incidence reported is a minimum because the number of females that are injured and die at sea is unknown and could be significant.

The ubiquitous mating urge of males, coupled with "unrequited love," leads some to direct their attention toward inappropriate mates, such as weaned pups (Rose et al. 1991). At a certain time of year, the latter are effectively "adult female substitutes." Males mount weanlings as they would an adult female, but because of the smaller size of the weanlings, the "neck bite" often becomes a "head bite" (Figure 6.3). This leads to severe injury and death. The circumstances in which this behavior is observed coincides with the absence of adult females as the breeding season winds down and the number of weaned pups increases. Weaned pups are chased out of the harems where they were born and move inland to the elevated sand dunes nearby. This appears to be an adaptation to reduce the probability of contact with libidinous subadult males. Weanling numbers in the sand dunes begin to increase in early February and peak in mid-March, when the last adult females go to sea. The weanlings gather in loose groups, or pods of up to 30 or more individuals, during the day and move into the shallow waters at night.

The culprits are primarily subadult males ranging in age from five to eight years; adult males were rarely involved. All of the males that mounted weanlings were usually unsuccessful in copulating with adult females. They mounted weanlings from February to late March. By late March, this behavior was no longer observed because the males departed the rookery to forage at sea. It was at this time that dead weanlings were observed in the dunes and counted.

Researchers observed males in the act of inadvertently killing weanlings by trying to mate with them. None of these attempts resulted in copulation. Nevertheless, up to 50% of the weaned pups on the Año Nuevo mainland rookery at the end of one breeding season showed signs of having been mounted by males: tooth marks and injuries or scars on the head and neck. During the years 1968–1990, 75 weanlings died on the Año Nuevo rookery. Most of them were in good condition before death. Necropsies revealed that at least 35% of them were killed by males during attempted copulation. All moribund weanlings showed evidence of neck bites, head injuries, blood oozing from the nose and mouth, or internal vascular trauma. It is notable that head injuries were especially likely at this time because the weanlings were only four to six weeks old and the sutures in the skull were not yet fully closed.

A few juveniles, two to three years old, were also killed by over-amorous males, but these are rare because juveniles and subadult males are rarely in residence at the same time.

Evidently, strong libido, coupled with sexual inexperience and limited access to estrous females, endangers the lives of pups, weanlings, and juveniles and stimulates males to generalize their sexual response from adult estrous females to inappropriate mate substitutes.

The obvious conclusion is that there is strong selection pressure in this species for males to mate, and to mate with as many females as possible. (As Dr. Tatiana (Judson 2002) says: "Evolution does not obey human notions of morality, nor is human morality a reflection of some natural laws. The deadly sins would be different if they mirrored evolutionary no-no's. Lust, for one, would be deemed a virtue; chastity would be deplored.") From the male perspective, the urge to mate is positive because it motivates males to try (Box 7.5). The problem is that relatively few males succeed in achieving this goal. The "mating club" in these animals has few members. Being a member of the club is analogous to a slot machine that only pays off with a jackpot. You win big or not all. It is not surprising that males will do whatever it takes to be a member of this club. The competition starts early. For example, when pups are weaned, as the mother departs for the sea, they move out of the harem because the nursing females in the harem see them as a threat to steal milk intended for their own pups. Milk thieves are bitten and may be severely injured. Nevertheless, some weanlings reenter the harem and attempt to steal milk from unwary mothers (Reiter et al. 1978). Some even succeed in getting adopted from a female that has lost her own pup; that is, this weanling gets the resources from a second mother and is called a "double-mother suckler." Males far outnumber females in doing this and they are more persistent in attempted thievery. The advantage for males is that larger size at weaning is positively correlated with large size in adulthood, and large size is associated with effective fighting and achieving high social rank and mating success. In contrast, the payoff in larger size is relatively trivial for females despite being exposed to the same risks, so they are less likely to take the risk of stealing milk. After all, all of the females will get impregnated and

Box 7.5 Dr. Tatiano Replies to Your Letters
Dear Dr. Tatiano*:

I'm a young male northern elephant seal about to be weaned. I want to grow up to be an alpha bull. I really want to lord it over all the other males and I think I would enjoy mating whenever I want. I know the odds are low but what can I do?

Ambitious in Baja

Dear junior,

You are right about the slim odds. Indeed, only about 8 out of 100 males born will live to breeding age and only 1–2 survive to prime age. And, of course, you must prevail over all the survivors. That being said, you have to be totally committed and give it all you've got starting now. For starters, whine and cajol mom to engorge you will all the milk she has. Never mind depleting her resources, which might reduce the birthweight of your brother or sister in mom's next-year pregnancy. You want to be huge at weaning because size early in life is positively associated with size when you grow up, and great size helps you win fights that will put you on the throne. Here is what you do. When mom leaves, deflated like a spent sausage, stay in the harem and use guile, stealth, and cunning to get yourself adopted by a young mom who has just lost her pup. The poor dear will be primed and ready to treat you as her lost beloved and inflate you with 55% fat-rich milk for another two to three weeks if you are lucky. You will be supersized, having feasted on two mothers. If you fail to get extra milk by getting adopted, sneak back into the harem and steal milk from sleeping or unwary mothers. Push their own pup out of the way but be careful. Those bitches will bite and shake you to pieces because you are stealing food from their darlings. You can lose an eye or get killed. You will do better at pilfering milk than girl pups your age, because nature, in its providence, delays the eruption of your canines, relative to hers, and you will be more persistent and less easily discouraged by being rebuffed.

Your teen years will be boring and there is not much to do except gorge feed and grow, grow, grow. You can get a big edge over others going through puberty so chow down when the growth spurt is greatest. At age 5 or 6, you will feel the urge to get it on with the hembras during that orgy called the breeding season. Be careful, as the older bulls are twice your size and dangerous. They will do everything they can to maim you and tame your lust. If you persist, at least do what oppressed males in other species do, pretend to be a female and when you make your move on an hembra be subtle. If you attract attention, you are dead meat. Sneaking pays off big time in a number of fishes and frogs. For example, the big males in bluegill sunfish defend territories that females visit to spawn. Small males that resemble females will even flirt with the big males. But when a female comes along, the bogus female joins in the courtship and releases his sperm when the big male releases his.

Now suppose you have survived to age 8 in good health, and you are well endowed, I mean in gross body size. Attend the annual orgy and check things out

Box 7.5 (*cont.*)

and exploit opportunities. You may get a few copulations. Most importantly, learn how to play the game and gear up for the big push when you reach prime years at age 11 or 12, if you are in the lucky 1% that live that long. One good year will get you about 100 assignations, which could result in siring 50–100 pups. Not bad for about three months of fighting and sex. A select few bulls – revered and worshipped in the elephant seal Hall of Fame – dominate mating for multiple years. I know one bull that dominated mating in the same harem for four years in a row! This is the stuff of legends, but you never know, records are made to be broken.

How to get that one good year? Arrive early in the breeding season, fight your way up to a rank among the top five males. Minimize fighting and energy expenditure and get by as much as possible on threats and bluffs. You are in for the long haul. Let the other top males do all the housecleaning and security. Make your move when the hembras start to come into heat. A good opportunity is when alpha and number two or three gets into a knock-down, drag-out fight. When the victor comes back into the harem bloodied and exhausted, this is your opportunity. Buckle up, challenge him, and kick his butt. Bite him on the butt as he yields and retreats. Make him remember who is boss. He who rules the harem at this time reigns for the entire breeding season. Big score.

How to manage the daily affairs of state? Sleep only when all others are asleep. Keep all other males away from the hembras by threat of violence, and fight only when you have to. Punish severely any male caught mounting or copulating with one of your hembras. Enjoy the charms of the hembras when the urge suits you. Be blessed with stupendous powers of recovery. Performance often suffers when the demand for service is high. When the last female leaves, catch up on your sleep for a couple of weeks before undertaking the long migration to the Aleutians for a well-deserved, round-the-clock, Lucullan feast that lasts for about two months. Fatten up for next season. Long live the king.

* This is a take-off, some would say a rip-off, on Olivia Judson's clever, entertaining, informative book, *Dr. Tatiana's Sex Advice to All Creation*, 2002, A Metropolitan/Owl Book, Henry Holt and Company, New York.

reproduce. From the communal group perspective, the losers in inter-male competition for social status cannot just turn off their sexual urges, and consequently, they are a dangerous element on the rookery.

In summary, males have short reproductive lives. They must survive to prime age to have a chance of breeding with multiple females. The optimal male strategy is do everything possible to dominate mating in a harem of females for one or a few years. The few that achieve this goal realize exceptional reproductive success, possibly siring over 200 pups in life.

8 Females

Designed to Reproduce

Notre félicité vient d'elle.

Racine

The first order of business for pregnant females at the start of the breeding season in early December is to find a safe place to give birth. Most pregnant females return to the same rookery, and often to the very same location on the rookery, year after year. It is prudent to do what worked previously. The males are already in residence and the early arriving, pregnant females are in no mood or condition to mate, but males accost them, nonetheless. It helps if females are joined quickly by other females and they group together. There is safety in numbers.

This is a dangerous situation for a female from the time she arrives on the rookery to when she returns to sea after giving birth, nursing her pup to weaning, and mating (Box 8.1). The difference in size between the sexes alone shows what she is up against. A male northern elephant seal may be 3–10 times heavier than the female with whom he mates (Deutsch et al. 1994). The largest male (2,265 kg) is 10 times heavier than the smallest, youngest, estrous female (229 kg). Usually, there is a three- to sixfold difference in size between the sexes. To put this in perspective, imagine what this would be like for a human coupling. A female weighing 50 kg (110 lb) would pair with a male weighing 150 kg (331 lb) or with a male weighing 300 kg (661 lb). At the extreme, the female would mate with a giant male weighing 500 kg (1,102 lb)! Obviously, female seals must be cautious in dealing with the opposite sex.

As pregnant females arrive and clump together, the most dominant male on the beach takes control of the group or "harem" and prevents other males from approaching the females. Being one of many females lowers a pregnant female's probability of being accosted and disturbed by any male. It may mean that she can be accosted by the alpha male only, and he is usually busy keeping competitors at bay or accosting other females. So, the harem is a relatively quiet place for a female and is far safer than being alone on the beach and exposed to several males, especially when she is pregnant or has just given birth and is in no mood for sex (Le Boeuf 1972, 1974; Le Boeuf et al. 1972; Reiter et al. 1981).

In addition, a female in a harem is more likely to be inseminated by the alpha male, as opposed to being inseminated by a peripheral male of undemonstrated fitness, a male that doesn't have the right stuff to become an alpha. It is advantageous for a

---

Box 8.1 Dr. Tatiano Replies to Your Letters

Dear Dr. Tatiano*:

I'm a nubile young female northern elephant seal giving birth for the first time. I was anticipating a sort of Arabian nights seraglio, but the harem is totally awful and not at all what I expected. All the females in this harem are bitchy and aggressive. They won't even allow me a reasonably quiet place to give birth to this pup, nurse it, and rendezvous with the alpha male that I am dying to meet. These mothers are vicious, and each one is only out for herself and her precious pup. And to make matters worse, I am constantly being hounded by these teenage males looking for love in the wrong place and at the wrong time. I am pregnant, duh! Please advise.

Alice in Mirounga-land

Dear Alice,

Welcome to the real world. Male fighting and bluster get press coverage but competition between females is equally vicious and intense, and age and size rule the day. So, my advice is grow up. Until you do, you will have to make do. The old females are large, and they rule in the harem. Get used to it. In wolves and Cape hunting dogs, young females are prevented from mating by the older dominant females. If young females sneak a copulation and get pregnant, their pups are killed by the dominant females. At least you get to give birth and try to bring it off. In the Ottoman Empire, which endured for 34 generations over 700 years, the sultan's mother, the *valide sultana*, ruled the harem with an iron fist; she supplied her son with multiple wives, concubines, and slaves. Indeed, for 150 years they ran the entire empire while the sultans dallied.

Learn to coexist with the dominant females and your harem sisters. Be smart: "wisdom is your destiny; ignorance is your doom" (Croutier 1989). Protect yourself and your pup. Good luck, for you will need it. Your lot will improve with age and experience. By the time you have given birth three times, you will be experienced, which will earn you position in the harem's ruling class.

What is the alternative? Don't join a harem? Forget it. You show up pregnant and you are relegated to the harem periphery or an isolated beach nearby, and the subadult males will pester to mate *ad infinitum*. Even with all of its pitfalls, the harem gives you some protection from the high-drive libido of the numerous Amboy dukes constantly patrolling the suburbs for naïve seals like you.

Find an isolated beach on some remote island to give birth? I know that the best sites are already taken, but striking out into the unknown is a longshot and few daring independent souls pull it off. It is the last resort. The stars have to be aligned just right for this strategy to work. The new site must be protected so your pup has a place to learn to swim and dive. Prince Charming must find you, so nature will take its course and you will give birth again next year. If you succeed, you will become a legend, the founder of a new colony.

* This is a take-off, some would say a rip-off, on Olivia Judson's clever, entertaining, informative book, *Dr. Tatiana's Sex Advice to All Creation*, 2002, A Metropolitan/Owl Book, Henry Holt and Company, New York.

female to have her male pup sired by an alpha because he is likely to inherit the traits of his dad that led to elevated dominance status. That is, she is being a good mother that "chooses" a good father, and she reaps the benefits of a more successful offspring. After all, about the only choice the female has in this situation is to choose a male for his genetic makeup because that is all he will provide to his offspring. Getting in a position to mate with the alpha male is probably the best way for a female to choose who will father her pup. As the harem grows, the oldest, largest females – who are dominant as a group to younger, smaller females – reside in or near the harem center, which is usually in close proximity to the alpha male. It is a relatively quiet place in the storm of activity that rages in and around the harem. The youngest females, smaller in size and of lower age-related rank, are relegated to the periphery of the harem, especially on the seaside, where disruption to nursing is frequent because of the high surf, high tidal conditions, and the constant attention females get from peripheral males attempting to copulate with them.

Females are considered adults when they give birth for the first time; this means that they copulated and were inseminated the previous year. For example, a female primiparous at age 4 became impregnated the previous year when she was three years old. Age at primiparity varied from 2 to 5 during the 1970s and 1980s in central California (Le Boeuf et al. 1972; Le Boeuf and Reiter 1988). Most females gave birth for the first time at age 4 (54%), followed by age 3 (36%), 5 (6%), and very rarely age 2 (1.5%). This means that the latter were impregnated the previous season as one-year-olds!

These relative frequencies of age at primiparity, however, vary with time and colony status; the frequency depends on whether the population is expanding or stable. Early primiparity is characteristic of expanding populations and later primi-parity for stable populations. The reason is that intra-female competition for breeding space is greatest when colony numbers are stable; in this situation, it pays to be large and older in order to successfully wean pups. Young primiparous females do best when the population is expanding and there is ample breeding space and less competition. This is well documented in southern elephant seals. At South Georgia during the 1950s, females gave birth for the first time at age 3 (Laws 1953); at other more stable Antarctic rookeries like Macquarie and Heard islands, primiparity occurred between the fourth and seventh year (Carrick et al. 1962).

Like males, females become reproductively active while still growing (Deutsch et al. 1994). Females giving birth for the first time at age 3 have attained only 75% of their mass at full maturity. Females stop growing in length and mass at about age 6 when their mass at parturition averages 541 ± 58 kg.

Virtually all females on the rookery during the breeding season deliver a single pup about six days after arrival (Figure 8.1). Twinning has not been observed in northern elephant seals but occurs rarely in southern elephant seals. The mother nurses the pup for four weeks. During the last one to four days of nursing, she is in estrus and may copulate a few times with one or more males. Of course, there are exceptions. Some females get more attention, or are especially accommodating, and may be in estrus for as long as 13 days, copulate at least once a day on as many as 5 separate days, copulate

Figure 8.1 From parturition to weaning. (a) Western gulls and other shore birds flock to gorge on the placenta as a female gives birth. (b) A female lies on her side and nurses her one-week-old pup. (c) A female nurses her pup that is about to be weaned. (d) Two recently weaned and newly molted one-month-old pups. (A black and white version of this figure will appear in some formats. For the color version, please refer to the plate section.)

as many as 13 times a day, copulate with as many as four different males, and copulate as often as 18 times throughout the entire estrous period.

After mating, the mother weans and abandons her pup and returns to sea to forage for about 2½ months. This is the end of her relationship with her offspring. She has put all of her maternal care into 26 days of nursing, during which she loses on average 169 kg or 36% of her initial body mass, measured just after giving birth (Deutsch et al. 1994; Crocker et al. 2001). Meanwhile, her pup gains 90 kg on average. The mean efficiency of mass transfer between mother and pup is 53%; that is, for every kilogram the mother loses, the pup gains about half a kilogram. All of the mass gain is from mother's milk, which is 55% fat in the late stages of lactation. For comparison, a good dairy cow's milk is 4% fat and human milk is 3%–4% fat. The milk of the hooded seal, *Cystophora cristata*, is 61% fat and contains the highest gross energy (5.9 kcal/g) of any mammal (Oftedal et al. 1988).

Older, larger females give more to their pups (Crocker et al. 2001), not just in mass transfer but also in milk energy. From a seal mother's perspective, weaning her pup in a healthy condition is the goal. She will not teach her pup how to swim, what to eat and where to find food, what predators to avoid that might eat them, or any other dangers they might encounter when they go to sea to forage for the first time. Weaning her pup in a healthy condition is enough; she has been a good mother. Now she is free to do other things without being encumbered. A mother and daughter may be on the same breeding beach at the same time in a few years, but they will not interact or shows signs of recognizing each other. Of course, she will not aid her daughter when she gives birth for the first time. She will not go into menopause and help any of her offspring breed or survive as many other animals do, for example, grandmothering in humans, elephants, and pilot whales (Hrdy 1999).

As described in Chapter 7 on male mating behavior, leaving the harem to go to sea and forage, after weaning their pups, is especially dangerous for a female because it exposes her to numerous peripheral males that are eager to mate with her. This can get violent and she can get mortally wounded (Mesnick and Le Boeuf 1991) (Figure 7.6). These males have been prevented from mating in the harem and their motivation to mate is high. This is their opportunity. The female must run a gauntlet of lusty males in order to reach open waters and begin feeding.

How does she do it? What does she do to defuse the danger? Most departing females assume the demeanor of being exceptionally receptive, proactive, and available to mate. Indeed, they spread their hind flippers to invite penile intromission in anticipation of a male's approach. This behavior is radically different from her response to courting males in the harem. Departing females that are proactively receptive receive fewer blows and injuries from suitors than females that resist. Thus, it seems that showing willingness to mate in this context helps the departing female "buy safe passage" to the water, which is precisely what she wants. After all, she has already mated in the harem and she is probably already inseminated. Her behavior is an effective means of reducing male aggression and the probability of injury; a male has no need to restrain a submissive female that assumes the position for mating. Moreover, if the most dominant of the peripheral males copulates with her, and he may do so several times, he often escorts her to deep water and defends her from other males. If the males that the female mated with in the harem were "rutted out" from excessive mating, and their sperm count was low, mating with males on her way out to sea gives her added insurance that she is inseminated. Nevertheless, some females are inadvertently killed by males when they depart harems to go to sea (Le Boeuf and Mesnick 1991).

Mating in the harem is more complicated because the female cycles from being pregnant to giving birth, to nursing, and then finally to being in estrus. Several males may try to mate with her regardless of her estrous condition. What do females do? They protest vocally and behaviorally whenever a male attempts to mount them; females do so even when they are in estrus. In one study (Cox and Le Boeuf 1977), a total of 1,478 mounts were observed, of which 79% were protested vocally by the female for the entire duration of each episode, 13.7% were partially protested, and 7.3% were not protested at all. Thus, protesting is not necessarily a sign of non-estrus; lack of protest or partial protest is a reliable indication that the female is in estrus. The word "protest" is used here as a term of convenience to describe the corpus of behavioral components that make it difficult for the male to obtain penile intromission. A female that protests issues a virtually continuous train of loud, rasping, vocal utterances that mean "no." This works to her advantage. The loud vocalizations alert all males within earshot that a female is being accosted. Her behavior activates the social hierarchy. It literally wakes up sleeping males and prompts them to live up to their social positions. They orient to the source of the vocalizations. They galumph directly to the pair. If the first male to arrive is dominant to the male trying to mate, he chases him off. He may attempt to mount the female himself, only to be chased off by another male that dominates him. The upshot of the female's attention-getting vocalization is that the only male that can continue to attempt to mate with impunity is the most dominant male in the area, the alpha male. He, alone, cannot be moved by others. A "reasoning" female might conclude that "if a male continues to try to mate with me, despite my protests, and he is not interrupted and chased off, he must be the alpha male. This must be Mr. Right, the best sire for my offspring." Indeed, under these circumstances, the female being mounted often stops protesting abruptly, acquiesces, and mates.

There is a logic to this argument. Females that discriminate between potential mates should be favored by selection because one-half of the genetic complement of their offspring comes from the father. That is, a female's reproductive success is linked to her offspring whose fitness is, in part, determined by an unrelated male. Because reproductive success is not simply a matter of the number of offspring that a female leaves but also a matter of the quality or probable success of these offspring, the male that is selected to be the other parent should be the most fit male available, the one whose genes will make the greatest contribution to the next generation relative to other available genes. When males invest nothing in the offspring beyond their sex cells, such as in elephant seals, females can only choose males on the basis of genotype. This is female choice in the broadest sense; it operates whenever a female influences what male will sire her offspring regardless of the means by which she brings this about. In the casual meetings of researchers in my lab, we spoke about the Okie bar effect in our own species. This is when a woman starts a fight in a bar and goes home with the winner. Why not? If you are a female elephant seal, this is how you get what works best for you.

The alternative to protesting by females is passive acceptance and facilitating mating (Le Boeuf 1972). This is most often observed during the latter stages of estrus (or during departure) and especially with a dominant male. A cooperative female responds to a male who puts a fore flipper over her back by lying still, elevating her perineum in the lordotic posture, and spreading her hind flippers. Insertion is signaled by an especially deep pelvic thrust that is held, resulting in an externally visible flexure of the sacral region of the male's spine. Full tumescence of the penis occurs with intromission. The male may pull the female toward him at approximately one-half minute intervals while at the same time pushing his pelvis forward. The female may elevate her perineum a few inches and move it slowly back and forth. Intromission, when uninterrupted, lasts about five minutes. Copulations in the water, although infrequent, last longer.

While mating is the be-all, end-all for a male, it is just one element in the reproductive life of a female. She is inseminated when she begins her 2½-month, post-partum foraging trip to acquire resources and recover the energy lost while nursing. But, technically, she is not pregnant yet. One of her eggs has been fertilized but implantation of the egg to the uterine wall does not occur until she returns to the rookery and finishes the one-month-long molt. Delayed implantation means that her gestation period lasts eight months, and this will enable her to give birth again at the same time of the year, and the next year, and the next, until she dies.

Another important element in the reproductive life of female elephant seals is provisioning offspring. Female elephant seals are capital breeders; they provision their pups using energy stores accumulated earlier, during the eight months at sea while gestating (Le Boeuf and Ortiz 1977; Riedman and Ortiz 1979; Ortiz et al. 1984; Costa et al. 1986; Crocker et al. 2001; Stephens et al. 2014). This is distinct from income breeders, like sea lions, that provision offspring using energy gained concurrently while nursing; that is, they nurse for a few days and then go to sea to feed for a few days and so on. All milk nourishment for elephant seal pups comes from the mother's body stores (blubber) accumulated in advance. The pup's birth and weaning mass reflects directly the nourishment received via the mother's stored energy determined by her foraging success during pregnancy. This is a closed system because the female elephant seal fasts while nursing and the pup gets nourishment from milk alone. Elephant seals have a highly efficient lactation system that is strongly impacted by body reserves that links foraging success at sea with reproductive success on land. It follows that age, maternal size, and body composition are important features of reproduction in northern elephant seals.

To understand how a population of animals adapts to the environment, and changes over time, it is necessary to track reproduction in females. This is more important than tracking mating success in males because, after all, all reproduction goes through females. Their success in producing progeny determines the genetic composition of the next generation and future generations. We expect that some females will do better at this than others, just as is the case with males. We want to know how variable females are in this respect. We want to know which females prevail in competition with others and why. To answer these questions, one must determine the lifetime reproductive success of individual females, which means monitoring the breeding behavior of females every time they breed throughout their entire lives. This is important but difficult and this is why long-term studies of lifetime reproductive success are rare.

Which females succeed on the reproductive battlefield? What do they do to achieve success? Why are some females better mothers than others? Which females produce the most offspring that are most likely to survive and breed? If a few males dominate mating, do some females dominate reproduction? After all, females compete with each other just as males do. It may not be as obvious as males winning bloody fights and culminating in mating with as many females as possible, but there is competition and it counts. The best metric for calculating female reproductive success is pups produced in life and, ultimately, the number of pups that survive to breed.

From the very start of our studies of the northern elephant seals at Año Nuevo in 1967, we kept track of the reproductive behavior of females tagged as weanlings, as well as some females tagged as adults. These individuals were identified with plastic alphanumeric tags and thereafter were identified annually, during the breeding season or the molt, by the name or number applied to the pelage with bleach or dye. Every breeding season, we recorded whether each identifiable female was present, gave birth, and nursed the pup to weaning or the pup died. We noted the sex of the pup, and in many cases, we weighed the pup at birth or, more frequently, at weaning. We called this housekeeping and referred to the dataset as the "parturition" record or the "known-age" or "known-history" record of females. We monitored this behavior annually during the breeding season and also identified the presence of these tagged individuals during the molt and the off season. We knew that collection of these data was of fundamental importance. Meanwhile, as we gave the highest priority to a variety of other studies of special interest over the years, the file on female histories grew increasingly larger. It was easy to postpone dealing with the database on females because of other pressing studies, and we rationalized that the larger the dataset, the more valuable it would become.

By 2018, the dataset on female histories consisted of over 8,000 individual female pups born in the years 1963–2008 whose behavior was recorded every breeding season of their lives from 1968 to 2018 (Le Boeuf et al. 2019). This is an interval of five decades that covered two to three generations of elephant seals. If an analogous study were done with human females, the data collection part alone would take 200 years. It was an enormous task. Hundreds of students and colleagues were involved in collecting data over the years, and the data had to be recorded, tabulated, managed, and synthesized. You might ask what took so long to do this study? Fifty-one years, or 55, depending on when you start the clock, is a long time. First, it takes a long time to collect the data, the entire lives of your subjects. Second, with a study like this, the larger the sample size, the more valid and reliable are the conclusions. So, there was merit in waiting. Lastly, given the time it took to do this study, it is not likely to be repeated, and therefore, it deserves to be treated in some detail.

The general conclusion of this long-term study is that great variation in lifetime reproductive success is characteristic of female northern elephant seals. There are huge individual differences in breeding success. This conclusion is supported by three categories of evidence:

(1) High mortality prior to breeding

Seventy-five percent of the weaned pups in the sample died before they produced pups. That is, only one of four pups weaned survived to breeding age. A steep decline in survivorship of this magnitude early in life is commonly seen in many mammals in nature (Wilson and Bossert 1971). I emphasize that this low survivorship is for pups weaned. Survivorship of pups born is even lower because mortality of suckling pups can vary from 8% to 70% per year, as revealed in studies during the 1970s and 1980s (Le Boeuf and Briggs 1977; Le Boeuf and Condit 1983). The exceptionally high values are associated with severe weather, coupled with high tides and surf washing over the breeding area at the peak season of births.

Most of the mortality of females prior to reaching breeding age occurs during foraging trips at sea (Le Boeuf et al. 1994). Surviving foraging trips during the first three years of life poses many dangers and is the greatest barrier to reproducing; 75% of the females weaned do not survive this period. The first foraging trip is especially dangerous because the pups are naïve about what to eat, how to get it, and who are the predators and how to avoid them. The first four to six foraging trips before females breed for the first time are discussed in Chapter 10.

(2) Low Reproductive Output of Most Young Breeders

Because young breeders, three to five years old, are the majority of females observed during any breeding season (because they haven't died yet), we expected them to produce the most pups and, as a group, show the highest lifetime reproductive success. (Figure 8.2) We expected them to have the greatest influence on the next generation. This was not the case. The majority of young female survivors died after producing only one or a few pups. For example, almost half of the survivors to breeding age (46% of them) bred up to age 6 and then died. Of these, 61% produced only one pup, 28% produced two, 9% produced three, and 1.4% produced four pups. Moreover, young females had a difficult time recovering from initial breeding events. Many females were not seen again after they bred for the first time: 8% for three-year-olds, 28% for four-year-olds, and 15% for five-year-olds. The lifespan of many females that bred early in life was shortened. This finding is not surprising. Females that breed at age 3, 4, or 5 are still growing (Reiter and Le Boeuf 1991; Deutsch et al. 1994). Breeding at this time means that resources must be channeled from growth to reproduction. This involves a cost that is reflected by failure to breed the next year, owing to insufficient recovery.

That's not all. Most young females giving birth for the first few times are poor mothers; many of them fail to wean their pups in a healthy condition (Figure 8.3). This is most obvious for primiparous three-year-old females. Success in weaning pups increased with age up to age 6. The weaning rate increased from a low of 60% in primiparous three-year-old females to 70% in four-year-olds, 82% in five-year-olds, and then 88% at age 6 or older (Table 8.1). After age 13, the percentage of pups weaned started to decline (Figure 8.3). Note that the mean lifespan of females increased with advancing age at primiparity and so did their lifetime weaning success (Table 8.1).

There are multiple reasons for the poor reproductive performance of young females. Maternal experience is important. Young females do stupid things like turning away from their pup as it is born and losing track of it and failing to bond with it from the very

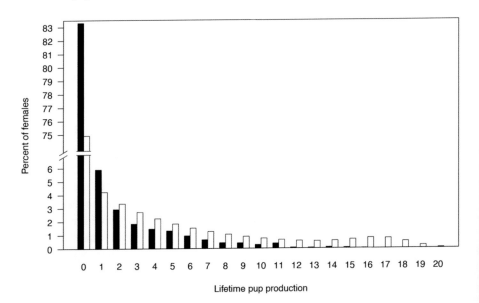

Lifetime pup production

Figure 8.2 The lifetime reproductive success of 7,735 female northern elephant seals as revealed by total pups produced in life. The black bars represent direct observations; the white bars are adjusted for tag loss, overlooked females, and females that emigrated. From Le Boeuf et al. (2019).

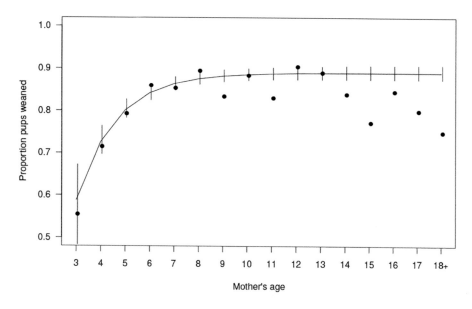

Figure 8.3 The proportion of pups weaned as a function of mother's age.

start. In addition, young females are often bullied away from their pups and isolated from them by larger, dominant females. Consequently, they are relegated to the periphery of harems where nursing is often disrupted by males attempting to mate with them. These encounters lead to mother–pup separation, the major cause of pup mortality on the rookery during the breeding season. Moreover, young mothers, being smaller, have less reserves from which to produce milk to feed their pups (Figure 8.4). Consequently, their pups are usually smaller than the pups of older, larger females and are less apt to survive and reproduce (Figure 8.5). Weaning mass of pups increased linearly from 96 kg for 3-year-old mothers to 133 kg for 7-year-old mothers and thereafter reached an asymptote of 138 kg for females between 8 and 17 years of age.

Pup weaning mass was associated with increased lifespan and lifetime reproductive success (Table 8.2). The lifespan of the heaviest weanlings was more than double that of the lightest weanlings, the probability of breeding increased threefold, and breeding appearances doubled. Pup weaning mass was associated positively with reproductive success as evidenced by a higher probability of breeding as an adult. For example, the probability of breeding was 36% for pups with a weaning weight of 150 kg or more, compared to only 12% for weanlings weighing 90 kg or less.

(3) Exceptional Reproductive Success of Long-Lived Females

The biggest surprise was that long-lived multiparous females were responsible for the most pups produced. Less than 1% of the females in the sample were exceptionally successful, producing up to 20 pups in life. Females that gave birth to 10 or more pups in their lives were responsible for 55% of the total pup produced. This stands out because these females were outnumbered three to one by females that produced fewer than nine pups in life.

Table 8.1. Lifetime breeding experience, lifespan, and weaning success as a function of age at primiparity in female northern elephant seals.

Age at primiparity (years)	N	Mean lifetime breeding appearances	Future expected lifespan (years)	Mean lifespan (years)	N	Weaning success at age				
						3	4	5	6+	Lifetime
3	184	3.8	4.16	7.16	313	0.6	0.7	0.82	0.88	0.79
4	649	4.17	4.55	8.55	1,537		0.72	0.82	0.87	0.83
5	210	3	3.07	8.07	313			0.72	0.87	0.83
6	104	2.93	3.12	9.12	116				0.89	0.89
Sum	1,147				2,279					

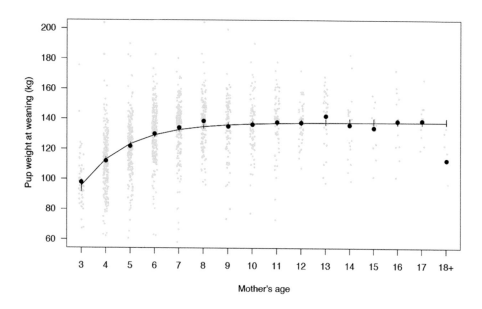

Figure 8.4 Pup weaning weight as a function of mother's age.

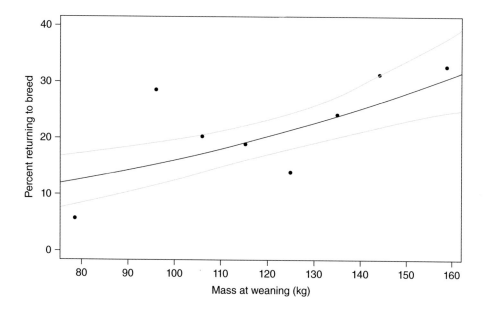

Figure 8.5 Percentage of pups returning to breed as a function of their mass at weaning.

The extraordinary performance of these females is amplified by further details. First of all, these females started breeding early, at age 3 or 4. As stated earlier, breeding early is a gamble, and most females lose, but great benefits accrue to those that pull it off successfully. Females that produced 10 or more pups in their lifetimes were the exception; they reproduced successfully early in life. They weaned pups successfully

Table 8.2. Lifespan, probability of future breeding, and breeding appearances as a function of mass at weaning in female northern elephant seals.

Mass at weaning			Future survival and breeding		
Mean (kg)	Range (kg)	Sample size (N)	Mean lifespan (years)	Probability of breeding (%)	Mean observed breeding appearances
79.37	<90	75	1.08	12	0.53
95.88	90–100	81	1.67	23	0.83
105.57	100–110	127	1.60	23	0.71
115.10	110–120	180	1.49	17	0.74
125.05	120–130	208	1.25	16	0.40
135.26	130–140	221	2.03	24	0.99
144.04	140–150	183	1.80	25	0.72
159.99	150+	103	2.86	36	1.09
Sum		1,180			

right from the start, having a 90% weaning success rate even as primiparous three- and four-year-olds. This is like investing money in the stock market early before others and reaping compound interest. In the case of seals, it is compound genetic interest that these females accrue. Starting early, these females bred consistently and for some virtually every year throughout their long lives. One female produced 20 pups in her lifetime. Another female produced a pup at age 22 and lived to age 23, the longest-lived female in the study. Another female produced pups for 16 consecutive years. Twelve females produced pups for 14 years, fifteen females for 13 years, and twenty-one females for 12 years.

To put the performance of these supermoms in perspective, it helps to categorize all the survivors that bred into parity groups (Table 8.3). By this, I mean parity group 1 consisted of all females in the sample that produced only one pup in life; parity group 2 consisted of females that produced two pups in life, and so on. An analysis such as this makes a number of points.

Parity group 0 was made up of 75% of the females in the sample; they produced no pups.

Parity group 1, consisting of 326 females, produced no more pups in their lives than 17 females in parity group 19. What makes this remarkable is that there were 20 times more females in parity group 1 than in parity group 19. In other words, 17 females that lived long did as well at producing pups as many more females that bred once and then died. Calculated another way, parity groups 16–20 produced more pups (3,076) than parity groups 1–5 (2,871 pups) despite being outnumbered by a ratio of six-to-one females. The mean number of pups per female was 17.2 in the old group and 2.6 in the young group.

Parity groups 16 and 17 produced the most total pups, 8% and 8.1% of the total pup production, respectively.

Table 8.3. Total pups produced by each parity group of females based on a sample of 7,735 weaned pups. Survivorship is corrected for tag loss based on a study of branded seals (Condit et al. 2014).

Parity group:		Group size:	
Lifetime number of pups produced	Number of females in group	% of females in sample	Total pups produced
0	5,794.01	74.91	0
1	326.40	4.22	326
2	259.21	3.35	518
3	210.20	2.72	631
4	172.00	2.22	688
5	141.53	1.83	708
6	117.03	1.51	702
7	97.30	1.26	681
8	81.33	1.05	651
9	68.32	0.88	615
10	57.70	0.75	577
11	49.19	0.64	541
12	43.06	0.56	517
13	40.97	0.53	533
14	44.59	0.58	624
15	52.62	0.68	789
16	60.82	0.79	973
17	58.21	0.75	990
18	40.04	0.52	721
19	17.04	0.22	324
20	3.40	0.04	68
Sum	7,735	100	12,176

Of the females that bred, the fewest pups were produced by parity groups 20, 19, and 1. Parity group 20 consisted of only three females and accounted for about 0.6% of the pups produced. After parity group 20, parity groups 19 and 1 produced the fewest pups, each group about 2.7% of the total pup production.

It is notable that total pup production decreased substantially in the oldest females, that is, a 55% decrease in pup production from parity group 18 to parity group 19 and a 79% decrease from parity group 19 to parity group 20. This indicates that senescence starts to rear its ugly head. The oldest females showed a decline in weaning rate as well as a reduction in the weight of their pups. Nevertheless, most of these long-lived females continued to breed. If they had a motto, it might be "bop till you drop," because they bred until they died. They showed no signs of menopause. This is in contrast to 14-year-old males who are still eager to mate but are excluded from doing so by the power structure.

The largest increase in pup production, 59%, was from the parity group 1 to parity group 2.

What is clear from this long-term study is that females that live long and produce pups annually are supermoms. They were successful in weaning their pups even during their initial breeding attempts, they bred annually with few or no skips, and they weaned their pups at a high rate. As elephant seals age, they get larger until growth plateaus in the later years. Larger females accumulate a larger fat store during gestation, and consequently, they have more milk energy to give their pups. Hence, their pups are larger at weaning, and this translates to increased lifespan and a higher probability of breeding. The females that managed this strategy successfully are the supermoms that determine the genetic composition of future generations.

What else does this long-term study reveal about the reproductive behavior of females? In most mammals females outlive males; the median lifespan of females is on average 18.6% longer than conspecific males; in humans the female advantage is on average 7.8% (Lemaitre et al. 2020). In elephant seals, the magnitude of the discrepancy in maximum lifespan is indicated in Table 8.3. If we assume that a female giving birth for the first time is four years old, we calculate that 361 females survived to age 14 or older, the maximum age that males attain; the maximum lifespan of females is 64% greater than that of males!

Annual fecundity – interpreted here as the percentage of parous females on the rookery during the breeding season – was very high, 97.5%, as reported in previous studies (Reiter et al. 1981; Le Boeuf and Reiter 1988). Fecundity would be lower if defined as the percentage of pups produced per opportunity in a population of individuals over their lifetimes and if there were skips in the breeding records. That is, the latter estimate of fecundity would be reduced by failure to breed in certain years. Skips might be due to failure to breed or poor coverage; that is, fecund females were present but not detected. We could not distinguish unequivocally between the two. Since the proportion of non-detected females increases with crowding and can be as high as 40%, we assume that most gaps in breeding records were due to non-detection of females with pups. Moreover, a recent report showed that 10% of females carrying satellite tags showed up not pregnant before or after the breeding season (Hückstädt et al. 2018). It is not known whether these females were fecund and aborted or whether they never got pregnant.

Nevertheless, one thing is clear. Skips in breeding records were absent in some supermoms; that is, they did not fail to breed in any years of their long lives. One supermom bred for 16 consecutive years, from age 4 to age 19; 21 other females bred consecutively for 11–15 consecutive years without skipping a breeding season. Among 77 females that gave birth to 10 or more pups, 49% had no gaps in their breeding records and 31% exhibited only one gap in a lifetime of breeding. Clearly, consistent breeding over a lifetime occurs in some females, indicative of elevated fecundity and high breeding success. The data also indicate that gaps in breeding records did not vary systematically with female age or cycling weather events such as an El Niño. In summary, this long-term study confirms that fecundity, using either definition, is between 87% and 97.5%.

This study also confirms reports of previous studies that juveniles disperse widely prior to breeding. Juveniles born at Año Nuevo may appear at nearby rookeries such as the Farallon Islands and Point Reyes Peninsula and mainland sites to the north (Le Boeuf et al. 1974; Allen et al. 1989). Approximately 10%–13.5% of females born at the Año Nuevo colony emigrate annually to these colonies to the north. It is notable that these emigrants visited the sites, "prospected," before they emigrated to breed there. Researchers at these colonies reported to us whether these females gave birth or whether they were just passing through.

The age structure of breeding females in the Año Nuevo colony ranged from 2 to 23 years. Four-year-old females were the most numerous. There were four times as many females breeding at age 4 as at age 3. After age 4, the percentage of breeding females declined steadily with age. Only one female bred at age 2.

This study showed that the prime reproductive years of females is from age 3 to about age 14, a span of 12 years. The proportion of pups weaned starts to decline in older females. Signs of senescence begin to appear at 15–18 years of age, as reflected by increased failure to wean pups and decreased weaning mass of pups. The latter suggests decreasing energy expenditure in the final years, but the change is small. The majority of females that live to age 18 or older, however, continued to breed until they died. Ninety-one percent of the oldest females bred one to four times after reaching age 18, and a 22-year-old female weaned her pup. This female returned for the next breeding season. We did not see her with a pup, but the observations were few. We are not certain that she was pregnant in her last year of life.

The sex ratio of pups born to 2,630 known-age females was biased to females early in life and males later in life, as theory predicts (Trivers and Willard 1973; Le Boeuf et al. 1989). The sex ratio increased from 48.9% males for three-year-old females to 55%–60% for females of age 10 or more. The difference is small but unlikely to be due to chance. According to sex ratio theory, and as applied to a polygynous species like elephant seals, it benefits a mother to have a large male pup when she is at prime age because size will benefit the male's chances of mating in adulthood. In contrast, mothers benefit from producing females when they are young and small, and still growing, because they produce small pups. In adult-hood, even small runty females will have no problem being mated, unlike the case with males.

We identified two supermoms that produced superdaughters. One of them, G7932, was born in 1983 and was observed breeding 10 times from age 4 to age 16. Her first born, GF24, developed into a supermom that bred at least 10 times from age 4 to age 15. The mother and daughter bred in the same breeding seasons on at least five occasions. In 1991, they both gave birth and reared their pups in harems separated by 150 m. They gave birth at about the same time of year each year that both of them were seen. The two of them were never observed in close contact with each other.

The observations of the other supermom that produced a superdaughter were similar. Both females bred in approximately the same location on the Año Nuevo mainland, but we never saw evidence that were together or recognized each other.

The two cases of supermoms producing superdaughters suggests that the attributes underlying superior reproductive performance might be, in part, inherited. It would take genetic studies, and much more work, to confirm that this is so or not.

The data collected from this long-term study are consistent with several theoretical predictions derived from previous studies of elephant seals, as well as other animals. Here are some of them:

- Large and experienced mothers are better mothers (Ralls 1976; Reiter et al. 1981; Sydeman et al. 1991).
- Large females wean larger pups that are most likely to survive and breed (Reiter and Le Boeuf 1991; Crocker et al. 2001).
- Breeding early in life reduces survivorship and fecundity in most individuals because there is a cost to breeding in individuals that are still growing (Williams 1966; Stearns 1976; Bell 1980; Reiter and Le Boeuf 1991; Sydeman et al. 1991; Sydeman and Nur 1994).
- Female elephant seals, like many mammals (Hrdy 1999), give birth annually until they die (Le Boeuf 1972; Reiter et al. 1981).
- Long lifespan of females is associated with greater reproductive output (Bercovitch and Berry 2017).
- Reproductive success increases with age up to a certain age (Gadgil and Bossert 1970; Pianka and Parker 1975) and then declines with advanced age and senescence (Promislow 1991; Gaillard et al. 1994; Nussey et al. 2013; Jones et al. 2014).
- Females bias the sex ratio of offspring, producing more females early in life and males later in life (Trivers and Willard 1973).

In summary, females that live long produce the most pups. The reproductive lifespan of a female is lengthened by early onset and extended duration. The oldest, largest females provide the most milk energy to their pups, which increases their odds of surviving and reproducing. To optimize reproduction, a female is pregnant or nursing a pup from the time she gives birth for the first time, throughout life, until she dies. As with males, a few long-lived females stand out as exceptional breeders. The difference in variance in lifetime reproductive success of males and females may not be as great as previously thought.

How do elephant seal mothers compare with other mammals and birds with respect to reproductive strategies?

We showed that age, lifespan, size, maternal experience, and energy transfer to pups are important variables leading to female reproductive success in elephant seals. There is abundant evidence that these variables have a profound influence on the reproductive success of females in numerous mammals and birds (Clutton-Brock 1988; Saether 1990; Fowler 1995).

Age and lifespan. As in elephant seals, age is important and long-lived females in many species are most successful. In African elephants, females that die before the median longevity of 40 years have reduced reproductive rates for their age while females who survived the mortality filter (Hawkes et al. 2012) sustain higher rates of reproduction throughout their lives (Lee et al. 2016). Similar results are reported in humans (Thomas et al. 2000), wild chimpanzees (Thompson et al. 2007), and a range of birds and mammals (Fowler 1995; McCleery et al. 2008; Rebke et al. 2010). Among reindeer (Weladji et al. 2010) and other ungulates (Loison et al. 1999;

Gaillard et al. 2000), survival and rates of reproduction are lower and more variable in young compared with prime-aged females. Social rank, which increases with age, is positively correlated with female reproductive success in spotted hyenas (*Crocuta crocuta*) (Holekamp et al. 1996) and mountain goats (*Oreamnos americanus*) (Côté and Festa-Bianchet 2001). Kid production increased with maternal age up to nine years and then senescence appears. Some old females continued to reproduce until they died. Similar results are reported for red deer, *Cervus elaphus* (Clutton-Brock et al. 1984, 1986), Barbary sheep, *Ammotragus lervioa* (Cassinello and Alados 1996), and feral ponies, *Equus caballus* (Selighsohn 1987). Reproductive success is strongly associated with increasing age in a variety of birds (Saether 1990). In contrast with wild animal life, human female life expectancy continues to increase into the seventh decade and beyond (Lemaitre et al. 2020). The unusual longevity in humans arises from reduced mortality throughout life, which is associated with better health care in modern times. With animals in nature, the healthy survive; the frail do not. What determines lifespan and life expectancy? Why do individuals differ in health and frailty and what is the role of the environment and genes in shaping this vital parameter of life?

Size. As with female elephant seals, size matters. Larger females in better conditions tend to invest more resources into reproduction than smaller females. Examples are African elephants (Lee et al. 2016), red deer (Clutton-Brock et al. 1982), sand lizards (*Lacerta agilis*) (Olsson and Shine 1996), Columbian ground squirrels (Broussard et al. 2005; Skibiel et al. 2009), black-tailed prairie dogs (Hoogland 1995), and several bird species (Price 1998).

Early breeding. As with elephant seal supermoms, females with the highest reproductive success tend to breed early in the reproductive period. Examples are red deer (Clutton-Brock et al. 1982), Richardson's ground squirrels (Dobson and Michener 1995), Uinta ground squirrels (Rieger 1996), European ground squirrels (Huber et al. 1999), and Columbian ground squirrels (Neuhaus 2000). Older birds lay earlier in the season, produce larger clutches or larger eggs, and provide superior coordination of parental duties (Fowler 1995). First reproduction, however, comes with a cost to many females as we have noted with northern elephant seal females; this effect has also been reported in female southern elephant seals (Desprez et al. 2014).

Energy transfer to young. As females of many species age they get larger, and this is associated with being capable of transferring more energy to offspring than smaller females. As in elephant seals, size of offspring at birth and weaning is positively correlated with survival and subsequent reproduction in many birds and mammals. Growth is prolonged in African elephants, and older, larger, and more experienced females allocate more milk energy to calves, thereby insuring offspring growth and survival. Female moose (*Alces alces*) increase reproductive effort with age (Ericsson et al. 2001). Quality of home range is most important for red deer hinds because this determines the energy transferred to calves, and older hinds invest more heavily in their offspring than young or middle-aged animals (Clutton-Brock et al. 1982). Offspring fitness is partly dependent on a female's ability to deliver adequate nutrients and energy through milk (Robbins et al. 1981).

Senescence. A decrease in breeding success of old mothers is characteristic of elephant seals as well as other mammals and birds (Reid et al. 2003; Nussey et al. 2006; McCleery et al. 2008; Sharp and Clutton-Brock 2010).

Social rank. Female elephant seals do not exhibit individual social rank, but age class dominance prevails. Older, larger females dominate younger, smaller females, and this alone is associated with older females having a higher weaning rate compared to younger females. In contrast, social rank is highly correlated with females' reproductive success in spotted hyenas (Holekamp et al. 1996). High-ranking females have preferred access to food, which transfers to the health and viability of their young. Moreover, these females prevent lower ranking females from mating. Social rank in hyenas is independent of size. In female mountain goats (Côté and Festa-Bianche 2001), kid production increases with age-specific social rank, and social rank and age are highly and positively correlated

Conclusion. It is clear from this brief review that female elephant seals and females of many mammals and birds have similar strategies for maximizing their reproductive success. Nevertheless, female mammals of different species may achieve reproductive success in several different ways (Holekamp et al. 1996). Several of the variables that influence female reproductive success – such as age, lifespan, size, and maternal experience – are conflated; they are inextricably linked together, and separating one from the other is difficult.

In species like elephant seals where females produce one offspring annually, nature selects females that breed early in life, breed at every opportunity, and live a long life. It selects large females with good resource-accruing ability, which results in large pups that are most likely to survive and reproduce. In comparison, nature does not select increasing life beyond 14 years in males of this species because they do not mate after 13 years of life. Moreover, in elephant seals, as in most mammals (Lemaitre et al. 2020), females outlive males. The discrepancy in lifespan between the sexes in elephant seals is large. If we assume that a female giving birth for the first time is four years old, we calculate from Table 8.1 that 361 females survived to age 14 or older and the oldest female lived to age 23. In five decades of studying elephant seals, we have never recorded a male that lived past age 14.

9   Diving, Foraging, and Migration

We are god-like; we who do things.

From Electra by Richard Strauss

When mating, birthing, and nursing are over for the year, adults of both sexes have lost a third to half of their arrival mass. It is time to go to sea to eat, replenish body stores expended, fatten up, and prepare for the next breeding season.

In the first half of the twentieth century, researchers studying pinnipeds viewed the sea as a black hole. It was a mysterious abyss that the seals entered to feed. Not much was known about what they ate, where they found food, how they captured prey, and virtually nothing was known about their diving and migratory behavior. The underwater life of marine mammals in their inhospitable environment was a mystery. All of that changed beginning in the 1960s with the development of a diving instrument, the time–depth recorder (Kooyman 1965, 1981; Gentry and Kooyman 1986).

The initial instrument was primitive, basically a kitchen timer and a bourdon tube for measuring pressure. The unit was attached to a Weddell seal, *Leptonychotes weddelli*, while it was lounging on the Antarctic fast ice near its diving hole that gave access to feeding in the water. The seal dove through the hole into the water, and when it resurfaced at the same hole, the only one available to it, the diving instrument was recovered. It recorded the depth and duration of the dive. Subsequently, the diving behavior of fur seals and sea lions were recorded. Females were good subjects too because they went to sea for a few days, carrying dive recorders, and had to return to nurse their pups, and the dive recorders with the data could be collected.

The time–depth recorder technology improved as quickly as Moore's Law; in a short time the analog instruments were replaced by microcomputers and sensors, and the instruments became considerably more sophisticated. Today, the diving instruments or biologgers attached to the animals are water-resistant miniature microprocessors, with multiple sensors, that measure depth and duration of dives, time at the surface, swim speed, flipper stroke frequency, acceleration, daily location to yield migratory tracks, water salinity, water temperature, angle of dive descent and ascent, heart rate, photos or videos of feeding or travel, jaw motion closure that indicates feeding, and more. Diving instruments attached to free-ranging elephant seals have also been used as oceanographic data collectors, taking the place of instruments that record temperature, conductivity, and depth, called CTDs (Boehlert et al. 2001).

Getting the measurements from instrumented seals is much cheaper provided that the oceanography of the northeastern Pacific, where the northern elephant seals go, is of interest. Archival instruments have sufficient memory to record for a year or more. The instruments are recovered when the animal returns to shore, and all data collected are downloaded to computers for analysis. Other instruments do not have to be recovered because they store data and then pop to the surface, exposing an antenna, and the collected data are uploaded to a satellite passing overhead in low orbit, from which the researcher downloads the data on his computer in the comfort of his home. This works well for whales where both attachment and recovery of tags is difficult.

The diving instruments are attached to the pelage of elephant seals with marine epoxy; the epoxy cradle falls off when the seals molt, exposing the new and unblemished pelage and skin (Figure 9.1). These onboard measuring devices are now being used on numerous marine mammals; the attachment varies with the species. The technology has revolutionized the study of the at-sea behavior of marine mammals to such a point that we now know more about the aquatic life of some pinnipeds than about their life on land. The black hole is being illuminated.

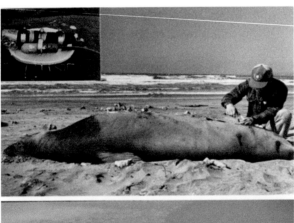

Figure 9.1 Dive recorders attached to elephant seals before they go to sea. Upper photo: attachment of the recorders with marine epoxy. The inset is a close-up of a time–depth recorder and a radio transmitter in the epoxy cradle. Lower photo: an instrumented yearling going to sea.

Elephant seals were considered good subjects for diving studies using time–depth recorders. This is because the units could be attached easily to the seal's fur while they were on land, and after being at sea for two to eight months, they returned reliably to the rookery. Analog dive recorders were attached to two lactating female elephant seals at Año Nuevo point on February 16, 1983 (Le Boeuf et al. 1986). Three days later, they went to sea for 93 and 127 days, respectively, before returning to the rookery, and the diving instruments were recovered. One instrument recorded the first 11 days at sea. The seal dived continuously, totaling 653 dives. On average, she dove 61 times per day, each dive lasting about 20 minutes followed by 3 minutes at the surface, resulting in her spending 89% of her time at sea underwater. The longest dive was 32 minutes. The average dive depth was 333 m, and the maximum depth was 630 m (Figure 9.2).

We were astonished! Our initial response was not to believe the instrument. Something must have gone wrong. This was the deepest dive recorded for any seal, at the time, and as deep as one estimated for a sperm whale. Ninety-four percent of all

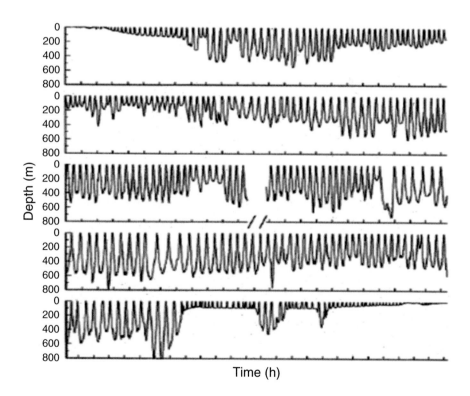

Figure 9.2 A portion of the two-dimensional diving record of an adult female showing depth of dives with time. The first seven hours show increasingly deeper dives as she crosses the continental shelf to deep waters. Diving is continuous and varies in depth from approximately 200–600 m per dive. There is a break in the record (slanted vertical lines) to save space. The record ends with increasingly shallow dives as the female crosses the continental shelf on her return to the rookery.

dives were deeper than 200 m. The continuous diving pattern was perplexing and unlike any other seal or whale. Did continuous diving reflect non-stop feeding? When did the seal sleep? How could the seal store sufficient oxygen to sustain a 30-minute dive? How did she recover from a dive so quickly? And so began the study of the diving and foraging behavior of elephant seals that continues to this day. Hindell et al. (1991) began their studies of southern elephant seals at Macquarie Island at about the same time. Much has been learned about the life of these seals in the black hole.

We humans take a deep breath before we dive, filling our lungs with oxygen to sustain us underwater for about a minute or two, if we are fit and have trained. Elephant seals exhale before diving because the lungs are not an important storage site for oxygen as they are with humans. Most of the oxygen of elephant seals (95%) is stored in blood and muscle. Blood oxygen storage capacity is determined by hemoglobin concentration and blood volume, and diving animals have higher hemoglobin concentrations and more blood than terrestrial animals (Box 9.1).

Box 9.1 MRI Scans of Elephant Seals!

MRIs are used worldwide to make detailed pictures inside human bodies to assess a variety of problems of the brain and spinal cord, the heart and blood vessels, the bones and joints, and many other organs. The MRI chamber is large and expensive. The technicians and radiologists who operate and interpret the scans can get a little bored doing the same thing every day. So, when we asked radiologists at the Lucas Center for Advanced MR Technology at Stanford if we might obtain MRI scans on elephant seals, they were intrigued and cooperative.

The idea was to examine blood flow during diving. The challenging and creative part, however, was getting the seal into the chamber and getting it to dive. We had a good idea of what organs to examine.

The subjects were four-month-old seals, capable of diving for 10 minutes or more, ready to embark on their first foraging trip in the open ocean. On the day of the experiment, the seal was placed in a conical nylon jacket and strapped to a restraining board in a prone position. The board was placed into a PVC half-pipe, which served as a fluid containment unit to prevent water damage to the magnet. (It was not necessary because the seals did not urinate *in situ*.)

Before imaging, the seal was fitted with a diving helmet made from a 35-cm Plexiglas tube, inner neoprene seal, and a secondary outer latex neck seal. The animal was allowed to acclimate for 30 minutes while a vacuum hose was attached to the helmet to ensure sufficient airflow. The seal took all of this in stride, relaxed, and slept.

When the experiment began, the helmet was flooded with cold water forcing the seal to "dive." The timing of the dive commenced when the animal's nostrils were completely submerged and continued until the helmet was drained and the first inspiration occurred. Each seal was subjected to four dives lasting five to seven minutes, slightly less long than the dives of freely diving pups. Readings were obtained before, during, and after dives.

Box 9.1 (*cont.*)

In all seals, the spleen, a known repository for red blood cells, contracted immediately upon facial submersion and hematocrit values increased and heart rate decreased. Concomitantly, the hepatic sinus increased in volume, indicating a direct shift of oxygenated blood from the spleen to the sphincter-controlled sinus of the venous system. We concluded that these two reservoirs act in concert to provide the diving seal with a means of red blood cell storage as well as a mechanism for controlled distribution of red blood cells into the general circulation, as needed, during the course of a dive (Thornton et al. 1997a, 1997b, 2001, 2005). This shift of blood from one organ to another is how a diving seal manages its oxygen, which permits it to dive long and forage efficiently.

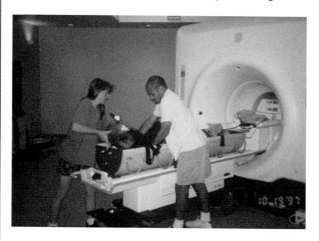

Figure B9.1.1 A juvenile northern elephant seal is placed in a holding module prior to being inserted into the MRI at the Lucas Center for Advanced MR Technology at Stanford University.

When an elephant seal dives its heart rate decreases, its oxygen consumption decreases, and the red blood cells in blood increase (Fedak et al. 1988; Andrews et al. 1997, 2000; Hindell and Lea 1998; Le Boeuf et al. 2000b). For example, the heart rate of free-ranging juveniles decreases from a mean of about 107 beats per minute at the surface between dives to a mean of 35 beats per minute during dives, with maximum decreases observed as low as 3 beats per minute (Figure 9.3). Blood flow to peripheral tissues is reduced and the lungs collapse at 25–50 m (Falke et al. (1985). In brief, the seals go hypometabolic while diving (Kooyman et al. 1973), especially during long migrations. Extreme hypometabolism allows elephant seals to spend up to 95% of their time at sea in breath-hold, exploiting depths down to 1.6 km for up to two hours from oxygen stores, that is, aerobically. Metabolism is lowered even more because the seal is operating in cold waters at depths. Indeed, it uses less energy diving

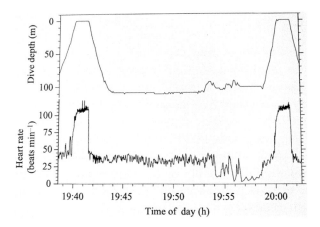

Figure 9.3 Dive depth and instantaneous heart rate from a juvenile northern elephant seal. Adapted from Andrews et al. (1997).

than resting at the surface. So, the elephant seal's secret ingredient is minimizing energy expended while maximizing energy intake foraging.

A great puzzle is how deep-diving elephant seals avoid the narcotic effect of extreme tissue nitrogen tension, oxygen poisoning, and other calamities (Miller 1972). Lung collapse at depth is one adaptation. The alveoli collapse first under pressure, and air is transferred into the cartilaginous airways where gas exchange is no longer possible. Consequently, nitrogen, which causes decompression sickness in humans, is mitigated. The seals, however, still have to manage high nitrogen loads and it is not clear how they do this.

Elephant seals spend most of the year in the water, hauling out on land once a year to breed and once a year to molt. Males of breeding age spend three months on the rookery during the breeding season, forage at sea for four months, return to the rookery to molt the skin and pelage, which lasts a month, and then go back to sea to forage. Thus, males spend eight months of the year at sea foraging. Breeding females spend a month on the rookery giving birth and nursing pups, followed by about two months at sea foraging, followed by one month on land molting, and then eight months at sea foraging while pregnant. Consequently, females spend more than ten months of the year foraging, two months more than males (Figure 5.1). Since males are significantly larger than females, they must consume more prey than females or feed on different prey that are more energy dense, and they have less time to do it.

When a seal leaves the rookery at Año Nuevo, it follows the bottom contour of the continental shelf that slopes gently to the open ocean (Le Boeuf et al. 1988) (Figure 9.2). The seal dives, swims along the bottom for 10–15 minutes, surfaces for a minute or so, then repeats the pattern until it reaches the continental slope and deep water in about 12 hours. Once in the open ocean, the seal continues to dive continuously, the dives get longer and deeper, the diving pattern changes, and it travels horizontally using a yo-yo diving pattern, which is repeated up and down diving in which each dive covers 500–1,300 m followed by 2–3 minutes at the surface. The seals do not swim at the surface. Individual males and females travel alone and take their own individual routes of travel.

When the seals reach deep water, the behavior of the sexes differs, as reflected by their diving behavior, speed of travel, migration routes, and foraging location. For this reason, we treat each sex separately. We restrict coverage in this chapter to adults. Diving of pups and juveniles is discussed in Chapter 10, on development.

Dive types. Dive recorders yield much information. A record of dive depth as a function of time reveals different dive shapes. Measurement of swim speed enables one to calculate descent and ascent rates and distance traveled per dive. Adding geolocation measurements obtained from changing light levels permits associating diving behavior with location. Later developments, such as three-dimensional movements and video and sound recordings, provided additional vital details of diving and foraging behavior (Box 9.2).

Diving records from elephant seals distinguished four fundamental dive types: transit, drifting, and pelagic and benthic foraging (Le Boeuf et al. 1988, 1992, 1993; Naito et al. 1989, 2013; Asaga et al. 1994; Crocker et al. 1994, 1997; Hassrick et al. 2007; Le Boeuf and Naito, in press). Similar dive types were described in southern elephant seals (Hindell et al. 1991; Jonker and Bester 2004; Bailleul et al. 2007; McIntyre et al. 2011). Dive types indicate the activity of the diving seal (Figures 9.4 and 9.5) and give us an idea of what the seals are doing.

Box 9.2 Elephant Seal Dives in 3D

A variety of sensor additions to data loggers attached to free-ranging elephant seals, such as video, swim speed indicators, accelerometers, triaxial acceleration, magnetometry sensors, GPS, and jaw motion recorders (Davis et al. 2001; Mitani et al. 2010; Naito et al. 2013; Adachi et al. 2017; Saijo et al. 2017), enable the construction of three-dimensional dive paths that provide insights into the details and function of dives as well as feeding behavior. For example, in "drift," also known as C dives, the negatively buoyant seal strokes during descent (C1) and then abruptly stops stroking at about 200 m and drifts down slowly for about half the duration of the entire dive duration (C2) before ascending to the surface (C3). During the C2 drifting phase, the seals rolled over and sank on their backs, wobbling periodically like a falling leaf. This enabled seals to decrease their descent rate while allowing them to rest, process food, or possibly sleep at a depth below surface predators. By adopting the belly-up position, they decrease the downward drift rate and maximize the time spent drifting and its benefits.

Jaw motion sensors, which reflect capture and consumption of prey, were operative soon after females departed the rookery, operated daily, and lasted until the end of the period at sea. An onboard camera indicated that the seals encountered, and presumably consumed, small mesopelagic fishes such as myctophids and a bathylagid. The seals foraged on prey items aggregated in patches, they consumed prey on 70%–90% of dives, and they used suction-feeding to capture prey.

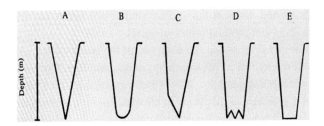

Figure 9.4 A schematic representation of the two-dimensional dive types observed in the diving records of northern elephant seals. The putative function of each dive type is: A and B – transit; C – digestion, rest, or sleep during the bottom drift segment of the dive; D – pelagic foraging; and E – benthic foraging.

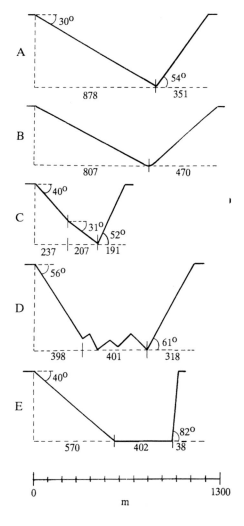

Figure 9.5 The dive types when the added dimension of speed and horizontal distance covered is added to depth. Adapted from Le Boeuf et al. (1992).

Transit dives are characterized by a shallow angle of descent to a pointed or rounded bottom followed by a shallow ascent to the surface; a single dive may cover 1.3 km horizontal distance traveled. Some of these dives are very deep, to 1,000 m, suggesting that the seal explores deeper depths while searching for prey. If it finds prey at depth, it will capture the prey. This is a strategy that females use to search for prey as they migrate in transit. If they find prey, the dive type changes from transit to pelagic foraging. Pelagic foraging dives are characterized by several vertical excursions at the bottom of dives, or "wiggles," indicative of pursuing and feeding on prey. Consumption of prey at the bottom of these dives has been confirmed by video and jaw motion closure (Naito et al. 2013, 2017; McGovern et al. 2019). Pelagic foraging dives show the deepest angles of descent and ascent. That is, the seals go directly down to forage and directly up to breath. Pelagic foraging dives show how females forage. Benthic foraging dives show how males forage. These dives have a flat bottom and are associated with foraging on or near the depth-limited continental slope. Drift dives are characterized by the seal stroking during descent to a depth, such as 200–400 m, where swimming stops and the seal begins passively drifting for about 50% of the dive duration and then powers up to the surface. The bottom portion of the dive, or the second segment, consists of drifting. Drift dives are associated with long bouts of foraging. They indicate internal processing; the seal is digesting food or getting rid of metabolites, resting, or sleeping. During the drifting phase the seal may drift down or up depending on its specific gravity. Indeed, as the seal forages and its fat-to-lean mass increases, it becomes positively buoyant. The changes in drift rate is a measure of foraging success (Bailleul et al. 2007). The horizontal distance covered in these dives is only about 600 m.

In summary, the sexes forage in different locations, on different prey, and using different behavioral strategies. Females forage on mesopelagic prey in the deep scattering layer, a variety of animals that migrate vertically with light, shallower and nearer the surface at night and deeper down during the day. The depth of dives of females reflects this diel pattern. Males forage benthically near the continental coasts, ostensibly on larger prey, and there is no diel pattern to their diving.

Males. From the rookery, males move directly north or northwest to forage at sites at the end of the journey along the continental margins of coastal Oregon, Washington, Canada, the Gulf of Alaska, or the western Aleutian Islands (Le Boeuf et al. 2000a)(Figures 9.6 and 9.7). The further a male travels to forage, the faster he travels to get there. The largest males undertake the longest migrations. Males that forage along the coasts of Oregon and Washington arrive at foraging sites in 9 and 12 days, respectively, while they take 38–50 days to reach foraging sites in the Aleutians Islands. Consequently, males that travel far have less time to forage than those that forage closer to the rookery. For example, males that forage in the Aleutians can spend only 26–44 days foraging while males that forage in Oregon and Washington can spend up to 89 and 77 days foraging,

The longest migration, from Año Nuevo to the Aleutian Islands, is 4,259 km. Males travel this distance in 42 days at an average rate of 101 km per day. Once they reach their destination, the focal foraging area (FFA), they slow down to less than 0.4 m per second, the stall speed of the instrument, as they begin foraging (Figure 9.8).

Figure 9.6 Migration routes of 22 adult males (red) and 17 adult females (yellow) during spring and fall migrations from Año Nuevo, California (black square), during the years 1995, 1996, and 1997. Adapted from Le Boeuf et al. (2000a). Figure created by Carl Haverl. (A black and white version of this figure will appear in some formats. For the color version, please refer to the plate section.)

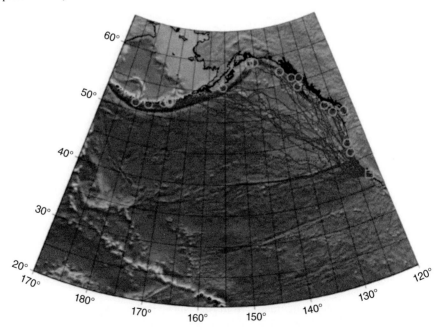

Figure 9.7 Satellite tracks of 22 adult males showing the outgoing leg (solid line) and return leg (dotted line) to the Año Nuevo rookery (open square). The final destinations, the focal foraging areas (FFAs), are shown as open circles (yellow). (A black and white version of this figure will appear in some formats. For the color version, please refer to the plate section.)

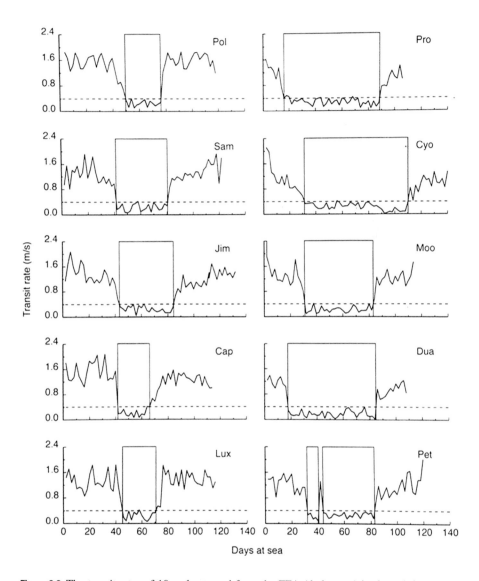

Figure 9.8 The transit rates of 10 males to and from the FFA (dark areas) is about 1.6 m per second. Transit is less than 0.4 m per second when males are on the FFA and are foraging.

This slow rate of movement is maintained all the while that the male is in the foraging area. The diving pattern to and from the foraging location is direct, consisting of 86% transit dives, which suggests strongly that males do not feed *en route*.

When males arrive at their destination, they gorge feed around the clock, exhibiting 73% benthic foraging dives interspersed with a few transit dives. The flat bottom of these dives suggests that they move over a relatively flat surface in search of prey or sit and wait for prey to come to them. Males that migrate the furthest, to the Aleutians, gain the most mass (34% over their departure mass or 12.8 kg per day), despite having less time to forage than other males. This suggests that the foraging at this location is better than at sites closer to home.

Each male forages in a single area that is relatively small and located near the continental shelf break. Individuals return to a specific site, such as an island, in successive years (Figure 9.9). Indeed, the migration tracks in successive years overlap to a remarkable degree. In the Aleutians, foraging sites were clustered around Amukta Pass, between Kanaga and Adak Islands and between Kiska and Amchitka Islands between 170° and 180°E.

The average duration of male dives is 22.3 minutes followed by 2.9 minutes at the surface. Male dive depths average 312 m, with maximum dive depths to about 800 m. The depth of male dives is consistent throughout the day and night, unlike that of female dives.

What do males eat? For starters, males are expected to consume about three times as much prey energy as females because they are 3 to 10 times larger than females. This may be achieved by more efficient prey capture, consumption of larger prey, or pursuit of different prey with a higher energy density. We know little about the diet of males.

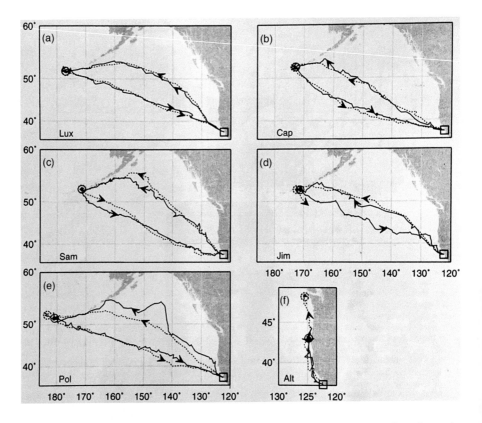

Figure 9.9 Tracks and FFAs (open circles) of six males from the Año Nuevo rookery (square) recorded in the fall (solid lines) and the subsequent spring (dotted lines) show that these males return to the same foraging site and take nearly the same routes during both migrations.

The few data available from stomach content analysis are biased to squid beaks and fish otoliths that pass slowly through seal stomachs and intestines. These data indicate that males eat skates, rays, ratfish, small sharks, cusk eels, and hagfish, and perhaps many other species found in the benthos (Hacker 1986). We know that adult males will eat and swallow meter-long dogfish sharks, *Squalus acanthis*, whole and headfirst because it has been observed and photographed (Albro 1980; see Figure 1 in Condit and Le Boeuf 1984). Consumption of rays is confirmed by the observation of stingray spines in the mouths of dead seals.

If males do not feed *en route* and restrict foraging exclusively to FFAs, as the diving pattern suggests, they alternate between extreme feasting and extreme fasting. Both the short period of gorge feeding and the length of fasting are remarkable. Here is a rough calculation for a male that breeds at Año Nuevo but forages in the Aleutians. He spends 90 days fasting during the breeding season, fasts for 42 days in transit to the foraging area, then forages for about 37.5 days, and then fasts for 40 days migrating back to the rookery and 30 days while molting on land. He goes to the foraging area twice a year. This adds up to 75% of the year fasting (290 days) and 25% of the year feeding (75 days). Note that a single fast may last almost half of the year or 175 days, that is, from the end of foraging in early October through the return to the rookery to breed for 100 days and then returning to the foraging area again in April. The pattern is to bulk up the blubber layer quickly twice a year and then "feed" off of it while fasting for the rest of the year. On the other side of the feast/famine equation, since a male loses about 40% of his mass during the breeding season (Deutsch et al. 1990), he must consume a considerable amount of prey that are energy-dense during the brief foraging period to replace the energy lost as well as to serve further growth, maintenance, and storage. In this respect, male elephant seals resemble the feast and fasting pattern of pythons (Secor and Diamond 1998; Secor 2009); they are record-setting, physiological curiosities!

Females. Females have a different foraging regime. When they leave the rookery and head straight across the continental shelf to reach deep water, they head northwest across a wide swath of the northeastern Pacific (Figure 9.6). They forage opportunistically along the way, searching for patches of prey that they exploit before moving on to search for the next prey patch. Generally, females do not move to a specific foraging area like males. Several areas of concentrated foraging, however, are apparent in the migration tracks of most females, and there may be significant overlap in tracks from year to year. The seals travel furthest from the rookery during the long foraging trip while pregnant, on average 3,257 km one way and over 9,850 km total distance traveled.

Considerably more research has been conducted on the diving behavior of females than males (Costa et al. 2012). Diving instruments have been deployed on 25 females or more for each biologger deployed on a male. Consequently, more is known about the at-sea behavior of females than males. The major reasons for this difference are that females are easier and safer to immobilize for attachment of diving instruments and they are more likely to survive and return to the rookery. One is more likely to recover the expensive instruments they carry than those of males. Another reason is that the behavior and reproduction of females are vital for monitoring the population.

Females average about 91% of the time at sea underwater. Percent time diving, as well as the dive rate, is similar in both sexes. Most females dive continuously throughout their time at sea; some females, however, show occasional extended surface intervals (ESIs), lasting a few minutes to up to a few hours. These ESIs and drift dives are associated with successful foraging. The average dive duration of females is 23 minutes followed by 2–3 minutes at the surface; the maximum dive duration is 109 minutes, that is, almost two hours. Dive durations and surface intervals are longer during pregnancy. Mean dive depth is 516 m, with a maximum dive depth of 1.7 km. Females dive deeper than males because they are feeding primarily in the open ocean while males are foraging along the depth-limited continental slope. A strong diel bimodal distribution of depth among females yields a daytime average dive depth of 619 m and a nighttime average dive depth of 456 m (Figure 9.10).

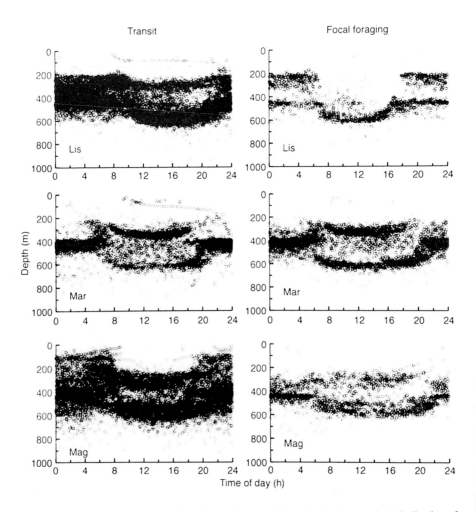

Figure 9.10 The dives of three adult females are deeper during the day than at night, indicative of foraging on mesopelagic prey that rise and fall with available light.

This indicates that females prey on inhabitants of the deep scattering layer, the dense aggregation of fishes, squids, and other micronekton (small organisms 2–20 cm in length) that move up and down daily with available light and provide an abundant, rich prey source for deep-diving predators like female elephant seals. Males do not show this pattern (Figure 9.11).

Females exhibit two types of dives most frequently: long strings of transit dives when moving and searching for prey and pelagic foraging dives when they have located a prey patch and are foraging. Feeding during the up and down movements at the bottom of pelagic foraging dives has been confirmed by jaw motion recorders attached to seals that record jaw closure and its association with the activity at the bottom of feeding dives (Naito et al. 2013)(Figure 9.12). Intense foraging over days or weeks is indicated by slow transit speeds or an elevated rate of drift dives (Crocker et al. 1997). Drift dives are most frequent during the long migrations when

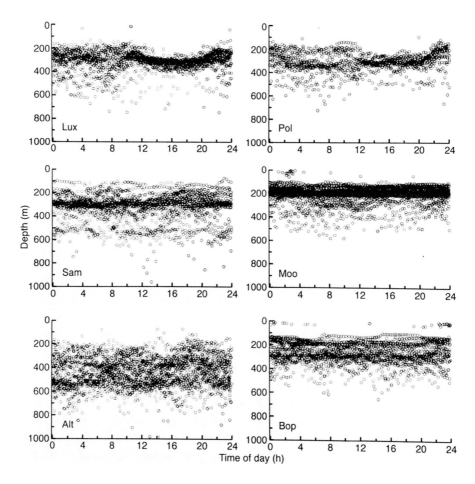

Figure 9.11 The depth of dives of six males while on their FFAs does not vary with time of day, indicating that males feed on different prey than females.

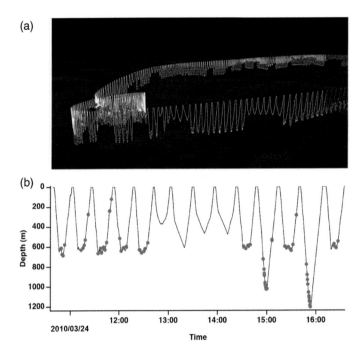

Figure 9.12 A jaw motion sensor attached to the lower jaw of a seal records when the jaw moves, indicating feeding. This figure shows the relationship between jaw movement events (JMEs) and dive depth and dive type from the diving record of an adult female northern elephant seal. (a) The three-dimensional distribution of JMEs (red dots) for the dives (white lines) along the migratory path of the seal in the northeastern Pacific. (b) An excerpt from the diving record showing JME distribution; dives 1 through 5 and 10, 12, and 14 are pelagic foraging dives; dives 7, 8, and 9 are drift dives (no feeding); dives 11 and 13 are transit dives; dive 6 is not classified. Adapted from Naito et al. (2013). (A black and white version of this figure will appear in some formats. For the color version, please refer to the plate section.)

the female is pregnant. Because females stop and forage many times along their migration routes, they travel less quickly than males, approximately 66 km per day during the post-breeding migration and 45 km per day during the post-molt migration when they are gestating.

Most females forage in the northeastern Pacific Ocean along a narrow band on the boundary between the subarctic and subtropical gyres (Le Boeuf et al. 2000a; Robinson et al. 2010, 2012). This is an area where waters of the cold nutrient-rich gyre mix with warm nutrient-poor waters of the subtropical gyre, which is optimal for the profusion of life. The majority of females forage in the mesopelagic zone at 200–1,000 m. Around 15%, however, exploit coastal and continental shelf areas during parts of their migrations. Here, some forage pelagically in deep water (Kienle 2019) while a few others may forage benthically part of the time, especially off the coast of Alaska and the eastern Aleutians Islands.

When females go to sea, after fasting on land, they are negatively buoyant and glide down on descent and stroke up on ascent, but this changes as the seal feeds and increases its fat content (its blubber layer) (Crocker et al. 1997; Aoki et al. 2011). At a certain point, the seal becomes neutrally buoyant and then positively buoyant and its strokes decrease in ascent. That is, there is an inverse relationship between buoyancy and the number of strokes required to surface. There is a parallel with drift dives. Females drift down when negatively buoyant and drift up when positively buoyant.

Prey/diet. In their distant past history, seals entered the sea for food. What are they eating in modern times? Early studies of the diet of northern elephant seals described the stomach contents of dead animals (e.g., Antonellis and Fiscus 1980; Condit and Le Boeuf 1984; Antonellis et al. 1987, 1994; DeLong and Stewart 1991). The general conclusion was that elephant seals consumed a variety of cephalopods, such as squids; teleosts, such as Pacific Hake; crustaceans, such as pleuroncodes; elasmobranchs, such as small sharks and egg cases; cyclostomes, such as hagfish and lamphreys; and tunicates, such as pyrosomes. These studies concluded that the primary prey was several species of cephalopods, many of which are luminescent. The method, however, reflected prey consumed during the last few days of transit, and because of the slow passage rate of squid beaks and fish otoliths in the stomachs of seals, bias was introduced in determining diet. According to this method, the diet of males and females were not significantly different.

Recent studies, using novel methods, shed new light on the prey of both species of elephant seals, and especially of females. Naito et al. (2013) used a head-mounted camera and jaw motion recorder attached to northern elephant seal females. Small mesopelagic prey (10–20 g), which included lantern fish (family Myctophidae), were consumed during diel foraging dives. In a subsequent study Naito et al. (2017) identified subadult or adult ragfish (*Icosteus aenigmaticus*) as prey. Consumption was confirmed from jaw motion recorders measuring jaw closure.

Goetsch et al. (2018) used quantitative fatty acid signature analysis and a library of prey fatty acid profiles from the mesopelagic eastern North Pacific, where female northern elephant seals forage, to estimate their diet composition. They concluded that mesopelagic fishes, such as myctophids, are the most important part of the diet of female northern elephant seals. The myctophids have high lipid content, making them one of the most energetically rich prey resources. They have an average energy content (gram/wet weight) up to an order of magnitude higher than that of squids. Viperfish and barracudina are also consumed but they provide half as much energy as myctophids. Most of these species are bioluminescent and are visible to foraging elephant seals. The fishes are ingested whole by suction.

Cues to the location of prey may be provided by visual sensing of bioluminescence; elephant seals have excellent vision (Levenson and Shusterman 1999). Vibrissae are used to detect the vortices created by the swimming prey (Miersch et al. 2011. Recent articles (Naito et al. 2017; McGovern et al. 2019) address seals searching for prey; video recordings show the vibrissae are erected and head movements accompany pursuit and prey capture. Prey capture is confirmed by audio sounds of swallowing or crunching.

Yoshino et al. (2020), using video cameras mounted on the head of female northern elephant seals, found that fish dominated the diet (78%) across all foraging locations, diving depths, and water temperatures, while squid comprised only 7% of the diet. Fish included myctophids, *Merluccius* species and *Icosteus aenignaticus*, and squids such as *Histioteuthis* species, *Octopoteuthis* species, and *Taningia danae*. These results align with the fatty acid analysis (Goetsch et al. 2018) and are in contrast with the stomach lavage analysis that concluded that cephalopods were the most important component of the diet (Antonellis et al. 1987, 1994).

Females capture 2%–8% of their body mass per day foraging (about 6.2% on average). This ingestion rate is the equivalent to 8–32 kg of prey captured per day spent foraging (20 kg on average). This is estimated to be 5–24 prey items per foraging dive. Seals that are initially fat acquire more prey energy, gain more weight, store more lipid, and birth larger pups. Mass gain of females averages 75.4 kg while foraging after breeding and 261.5 kg after the molting migration. The largest seals expend the least amount of energy on a mass-specific basis during the longer foraging migrations. This is reflected by flipper-stroking. Seals flipper stroke less during long rather than short migrations (24% lower). The largest pregnant seals stroke the least amount.

One researcher calculated that a single female elephant seal would consume about 5,600 kg of squid or 1,600 kg of fish per year from the North Pacific mesopelagic ecosystem. Taking this further, he estimated that all females in the population in 2010 were likely to have extracted 286 million kilograms of squid or 82 million kilograms of fish in a year.

Studies of southern elephant seals from the sub-Antarctic revealed a generalist diet (Hindell et al. 1991; Daneri et al. 2000; Piatkowski et al. 2002; Bradshaw et al. 2003; Field et al. 2005, 2007; Lewis et al. 2006; Campagna et al. 2007; Ducatez et al. 2008; Newland et al. 2009; Biuw et al. 2010; Eder et al. 2010). Females fed in deep waters on diel migrating mesopelagic prey. Males fed primarily on demersal and benthic prey (Hindell et al. 1991; Fedak et al. 1997; Campagna et al. 1998; Field et al. 2001; Eder et al. 2010; McIntyre et al. 2011; Hückstädt et al. 2012). The diet was comprised primarily of fish and cephalopods (Green and Burton 1993; Daneri and Carlini 2002; Lewis et al. 2006; Newland et al. 2009; Eder et al. 2010).

McGovern et al. (2019) identified the prey of female southern elephant seals from a head-mounted video camera as lanternfishes (*Lampadena* species) from the family Myctophidae and pencil smelts (*Nansenia* species) from the family Microstomatidae. Other prey identified was barbeled dragonfishes from the family Stomiidae. Most of the feeding occurred during pelagic foraging dives (97.5%); 0.8% occurred during drift dives and 1.7% occurred during transit dives. The number of prey encountered was 3.4 per foraging dive.

Little is known about the diet of male elephant seals of both species, most of whom feed benthically most of the time (Hindell et al. 1991; Campagna et al. 1999; Le Boeuf et al. 2000a; Simmons et al. 2007; Biuw et al. 2010). Males from Marion Island are an exception as they feed pelagically in deep waters near the island (McIntyre et al. 2012). We know that male northern elephant seals consume stingrays and small sharks (Albro 1980; Condit and Le Boeuf 1984) because it has been observed in nature and in captivity (BL, personal observation).

Evidently, northern elephant seals are not food limited in most years because most of them gain mass predictably during their lives. In addition, their prey is not sought, or not easily obtained, by human fishers. This was so during the population recovery over the last 130 years as well as now. Their remarkable recovery from near extinction was made possible, in part, because of the ready availability of food that only they could exploit. In a severe El Niño year such as 1997–1998, it is a different matter. In that year, some females spent more time foraging during the post-breeding trip to sea and gained little mass or lost mass (Le Boeuf and Crocker 2005; Crocker et al. 2006). Climate change, with the increasing warming of the seas, will make matters worse (McIntyre et al. 2011). It has been suggested that declines in some southern elephant seal populations, such as that at Marion Island, were due to a decrease in female survival owing to food limitation (Pistorius et al. 1999, 2004).

Navigation. It is obvious from the free-ranging migrations of elephant seals that they are exceptional navigators. After foraging up to 4,500 km from their colony, the seals return to their particular colony unerringly, avoiding the dozen or more other colonies along the 1,000 km of the coastline from central California to mid-Baja California, Mexico. Adult males return to the same FFA in the Aleutian Islands in successive years; the two migratory tracks to and from the islands are so similar, one fits closely over the other (Figure 9.9) (Le Boeuf et al. 2000a).

Elephant seals navigate well but we don't know how they do it. With free-ranging animals in nature, one cannot control the relevant variables – such as vision, acoustic signals, or magnetic fields. The mechanisms that explain the process of navigation are best studied in a laboratory setting, such as birds in a planetarium or honeybees in small enclosures.

Matsumura et al. (2011), however, used an intermediate method, the homing paradigm (Le Boeuf 1994; Oliver et al. 1998), to investigate homing (Box 9.3). The paradigm is analogous to studying navigation in homing pigeons (Vyssotski et al. 2009). Three juveniles were translocated from the Año Nuevo rookery and released at sea 60 km southwest and beyond the continental shelf in deep water. The three seals returned with a mean transit time of 2.2 days. Three-dimensional movements and heading directions underwater and at the surface were made from a triaxial acceleration and magnetometry data logger and a GPS tag attached to the seals. Swimming speed underwater was 1 m per second. The seals departed in different directions after release but two of them aligned with the colony soon thereafter. Subsequently, all seals traveled toward the coast, south of the home colony. After reaching the coast, they all traveled north to the home colony. Drift dives (Le Boeuf et al. 1996; Crocker et al. 1997) were observed in two of the seals. During the middle segment of these dives, the seals drifted passively spiraling around as they sank down. Despite over 20 complete spirals during the drift phase, they maintained their initial direction when they rose to the surface. That is, the seals appeared to maintain a directional sense while submerged, even while drifting and spiraling around. Evidently, the navigation cue is accessible and perceivable at deep depths away from surface cues. The cue may be the ambient acoustic field. Harbor seals are good at this (Bodson et al. 2006). The authors concluded that at the surface visual scanning of coastal landmarks aided

---

Box 9.3 The Homing Experiment

Homing behavior is widespread in animals (Papi 1992). Sea turtles, birds, numerous mammals, and California sea lions do it. Why not elephant seals? Since elephant seals spend so much time at sea, it is difficult to conduct short-term experiments lasting a day or two. With this practical application in mind, we asked if yearlings translocated from the Año Nuevo mainland would return "home" quickly and reliably. Of 120 yearlings translocated to various terrestrial or aquatic sites up to 100 km away from the collection beach in the spring or fall – when they were molting or resting – 88% returned home within 4½ days (Le Boeuf 1994; Oliver et al. 1998).

This translocation paradigm allowed us to conduct several short-term experiments such as measurement of swim speed, locomotor energetics from video images, heart rate and EKG during diving, metabolic studies with doubly labeled water, acoustic stimuli impinging on the diving seal, and whether the seals avoided or were disturbed by loud, low-frequency sounds (Burgess et al. 1996; Costa et al. 1996; Fletcher et al. 1996; Andrews et al. 1997; Davis et al. 2001).

Our nerves were frayed, and our blood pressure escalated, when we put an instrument package costing over $5,000 on a seal and released it to free range at sea. When Randall Davis attached a GPS antenna, a satellite telemeter, and a very expensive custom-made video system on a 27-month-old seal and released it 30 km south of Monterey Bay, he had to wait a long, nervous week for the seal and to recover the instrument package. He was able to construct the seal's three-dimensional dive path back to the rookery, which revealed that long periods were spent gliding during descent, and the seal took a direct path home, suggesting exceptional navigation ability. In one video recording, with the camera pointed toward the rear to record the sculling action of the hind flippers, we see the yearling being followed over the continental shelf by a curious harbor seal.

---

navigation, as has been shown for avian navigation (Vyssotski et al. 2009). Geomagnetic and acoustic cues are potential navigation cues because they are available underwater. It is likely that that the seals navigate using a variety of cues as do many migratory species (Quinn and Branton 1982; Maaswinkel and Whishaw 1999). Further research is needed.

In conclusion, males and females of both species are veritable diving machines that can manage a limited oxygen supply for up to an hour and a half while foraging and can withstand high pressure and changes in high pressure that would kill most mammals.

Figure 1.1 An adult male surrounded by pregnant and nursing females emits a threat vocalization to a competitor from the stereotypical posture. Note that he is squashing a pup that is crying out in distress. (A black and white version of this figure will appear in some formats.)

Figure 1.2 Males and females are easily distinguished from facial features or size. (A black and white version of this figure will appear in some formats.)

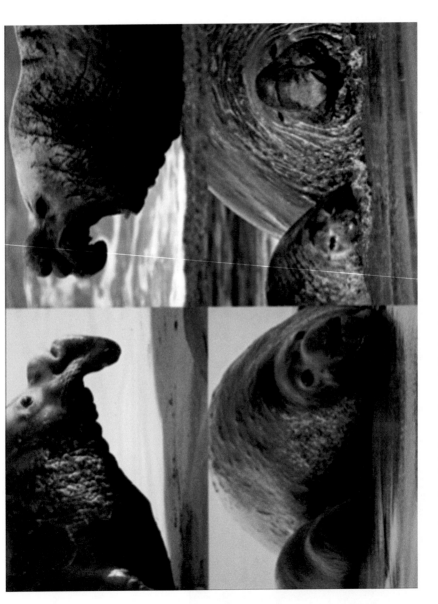

Figure 1.3 Both species of elephant seals are similar in appearance. Top: headshots of an adult male northern elephant seal on the left and an adult male southern elephant seal on the right. Bottom: adult female and male northern elephant seals on the left and adult female and male southern elephant seals on the right. (A black and white version of this figure will appear in some formats.)

Figure 2.1 Three elephant seal colonies. (a) Pilot Rock Beach on the northeast point of Isla de Guadalupe, Mexico. The beach is crowded with elephant seals at the peak of the breeding season. (b) A view of Elephant Seal Beach on the northwest side of Isla de Guadalupe. Access to this major breeding beach is difficult. (c) Año Nuevo Island and mainland. (d) The breeding area for elephant seals at Piedras Blancas in southern California. Courtesy of Friends of the Elephant Seals. (A black and white version of this figure will appear in some formats.)

(a)

(b)

Figure 3.3 Places to view elephant seals. (a) Visitors on a guided tour view the elephant seals at Año Nuevo from a distance. (b) A view of a portion of the elephant seal rookery at Piedras Blancas during the breeding season. The perspective is viewing south; Highway 1 is to the left of the photograph. (A black and white version of this figure will appear in some formats.)

Figure B4.2.1 Elephant seals can resemble dead seals, smooth rocks, or boulders, or even mountains. (A black and white version of this figure will appear in some formats.)

Figure 4.2 Santa Barbara Channel oil spill on March 17, 1969. (a) Northwest Cove on the western tip of Point Bennett, San Miguel Island, showing the crude oil that washed ashore in mid-March 2017 and blackened beaches and soiled weaned elephant seal pups. (b) A close-up view of weaned elephant seals enmired in crude oil at Northwest Cove. (c) Dick Peterson attaches a tag to the hind flippers of a weaned elephant seal pup with 75% of its body covered in crude oil. (A black and white version of this figure will appear in some formats.)

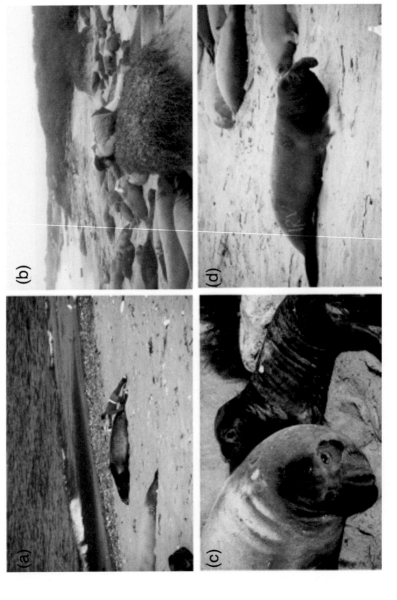

Figure 4.3 Marking seals. (a) Marking a sleeping adult male with a bleaching solution. (b) Marking an adult female in a crowd of females. (c) A mother and her pup marked with identical paint spots. (d) A subadult male, Irv, marked with the bleaching solution. (A black and white version of this figure will appear in some formats.)

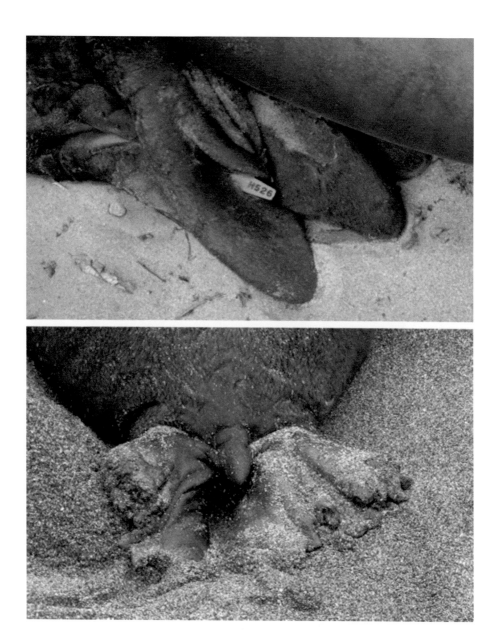

Figure 5.4 The hind flippers of a newly weaned pup showing that it was born at the Año Nuevo colony (top). The hind flippers of a newly weaned elephant seal (bottom) that have been bitten and chewed off by coyotes at the Point Reyes colony. Courtesy of Sarah Codde; Photograph by Marjorie Cox. (A black and white version of this figure will appear in some formats.)

Figure 5.5 Sand-flipping helps seals expose cooler sand below the surface and control their temperature on a hot day at Pilot Rock Beach on Isla de Guadalupe. Sand-flipping is also indicative of distress or nervousness. (A black and white version of this figure will appear in some formats.)

Figure 6.2 Nursing females. (a) A nursing mother, separated from her pup by two females, calls to it. (b) This female allows four pups to suckle, one of which may be her pup. (A black and white version of this figure will appear in some formats.)

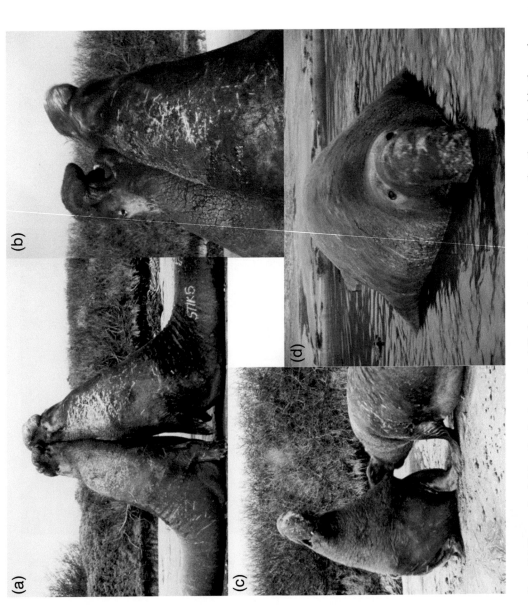

Figure 7.1 Two males fighting for dominance. (a–c) Each male attempts to tower over the other and to bite the neck or nose of the other. (d) The winner of the fight, shown here, is as bloody as the loser. (A black and white version of this figure will appear in some formats.)

Figure B7.4.1 Weighing males. (a) A subadult male is lured onto the scale by a decoy female emitting protest calls that she is being mounted. (b) An adult male is discouraged from moving further as he pursues the decoy and rests momentarily on the scale. (c) An adult male pauses on the scale. (d) The male's weight is recorded remotely as he rests on the scale. (A black and white version of this figure will appear in some formats.)

Figure 7.6 Four peripheral males chase a female (black arrow) as she departs the harem. The oldest, largest male is in position to block her escape and mate with her. (A black and white version of this figure will appear in some formats.)

Figure 7.7 A female with a bloody head wound caused by a peripheral male attempting to mate with her as she left the harem to go to sea. (A black and white version of this figure will appear in some formats.)

Figure 8.1 From parturition to weaning. (a) Western gulls and other shore birds flock to gorge on the placenta as a female gives birth. (b) A female lies on her side and nurses her one-week-old pup. (c) A female nurses her pup that is about to be weaned. (d) Two recently weaned and newly molted one-month-old pups. (A black and white version of this figure will appear in some formats.)

Figure 9.6 Migration routes of 22 adult males (red) and 17 adult females (yellow) during spring and fall migrations from Año Nuevo, California (black square), during the years 1995, 1996, and 1997. Adapted from Le Boeuf et al. (2000a). Figure created by Carl Haverl. (A black and white version of this figure will appear in some formats.)

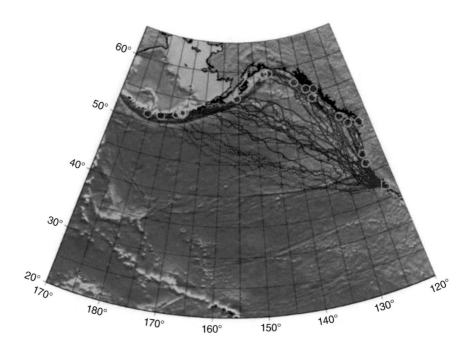

Figure 9.7 Satellite tracks of 22 adult males showing the outgoing leg (solid line) and return leg (dotted line) to the Año Nuevo rookery (open square). The final destinations, the focal foraging areas (FFAs), are shown as open circles (yellow). (A black and white version of this figure will appear in some formats.)

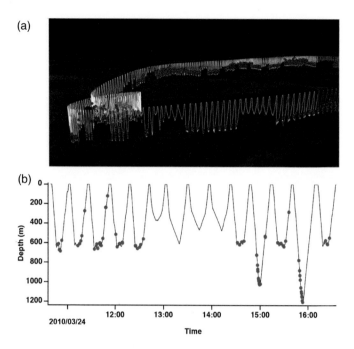

Figure 9.12 A jaw motion sensor attached to the lower jaw of a seal records when the jaw moves, indicating feeding. This figure shows the relationship between jaw movement events (JMEs) and dive depth and dive type from the diving record of an adult female northern elephant seal. (a) The three-dimensional distribution of JMEs (red dots) for the dives (white lines) along the migratory path of the seal in the northeastern Pacific. (b) An excerpt from the diving record showing JME distribution; dives 1 through 5 and 10, 12, and 14 are pelagic foraging dives; dives 7, 8, and 9 are drift dives (no feeding); dives 11 and 13 are transit dives; dive 6 is not classified. Adapted from Naito et al. (2013). (A black and white version of this figure will appear in some formats.)

Figure 11.1 An adult male in deep sleep away from the competition in harems. (A black and white version of this figure will appear in some formats.)

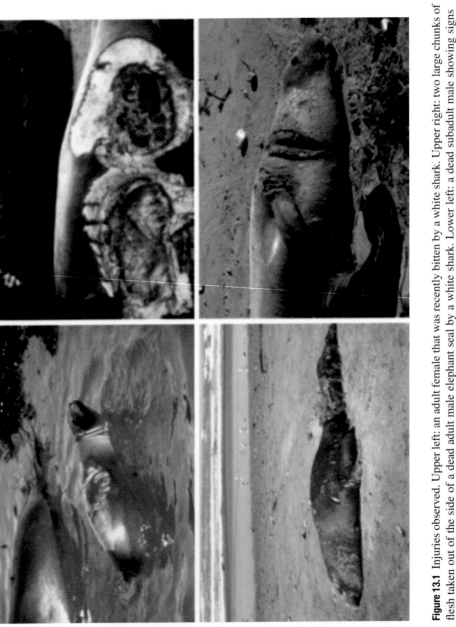

Figure 13.1 Injuries observed. Upper left: an adult female that was recently bitten by a white shark. Upper right: two large chunks of flesh taken out of the side of a dead adult male elephant seal by a white shark. Lower left: a dead subadult male showing signs of having been attacked and killed by one or more sharks. Lower right: an adult female with pup with a large wound on her neck and over her shoulder that appears to have been caused by a ship's propeller. (A black and white version of this figure will appear in some formats.)

10 Development

Don't criticize what you can't understand, your sons and your daughters are beyond your command.

Bob Dylan

The early development of elephant seals fascinates us because pups learn nothing from their parents, growing males and females must adopt different lifelong reproductive strategies, and the young must learn on their own to feed at sea and become prodigious deep and long-duration divers. How do they get this way? What is the transition like?

A starting point is the termination of parental investment in offspring, which varies widely in mammals. In most species, the transition from dependence on the parents for nourishment and protection to independence is gradual. Long after weaning from mother's milk, parental investment in the young may continue in the form of feeding solid foods, guarding, aiding, carrying, and training. This is the typical pattern in some rodents and ruminants, many carnivores, and several primates. At the opposite end of this continuum are a few species in which no further parental investment is provided after nursing ceases. Elephant seals are one of these species. A good mother elephant seal weans her pup with a large fat store. That is it. She abandons her pup. She does nothing more to ensure its survival and the passing on of her genes.

Pups that survive to weaning face a rough introduction to life. Abandoned by their mothers, they are driven out of the harem by aggressive females still nursing and guarding their milk supply for their own pups. This chapter treats development from weaning to the first trip to sea at 3½ months of age and then the next four trips to sea until they reach three years of age (Le Boeuf et al. 1972, 1989, 1994, 1996, 2019; Le Boeuf and Briggs 1977; Le Boeuf and Ortiz 1977; Ortiz et al. 1978; Reiter et al. 1978; Riedman and Le Boeuf 1982; Ortiz et al. 1984; Costa et al. 1986; Rose et al. 1991; Blackwell and Le Boeuf 1993; Le Boeuf 1994; Thorson and Le Boeuf 1994; Aurioles et al. 2006; Riofrio-Lazo et al. 2012).

The period immediately following the cessation of parental investment is a critical time in the development of mammals, especially for those that are cut off abruptly from mother's milk. The weanling is vulnerable to predators and must secure its own food at a time when it is relatively defenseless, inexperienced at survival, and not fully developed. During the 2½ month period after weaning, elephant seal pups fast, metabolizing water and nourishment from their blubber, and they undergo the behavioral and physiological changes that prepare them to make a living at sea.

Mothers wean their pups when they appear to us as "ludicrously obese" and "nearly globular" (Bartholomew 1952) (Figures 10.1 and 10.2). Their huge fat store allows them to fast for 2–2½ months while learning to swim, dive, and negotiate a new medium, the water, before going to sea for the first time in search of food.

After being chased out of the harems, weanlings congregate in the dunes above the beaches (Figure 10.3). Some perfectly healthy weanlings return into the harem and attempt to steal milk from nursing females (Reiter et al. 1978). This is dangerous because they risk serious injury. Most of them are bitten and rebuffed but a few manage to steal milk successfully or, best of all, get adopted and suckle several days or weeks from an alien mother who has lost her own pup. These "double mother" sucklers become oversized superweaners that may weigh almost twice the size of

Figure 10.1 A newly weaned and molted pup next to a pup that is a week old.

Figure 10.2 A mother (background) and her newly weaned pup.

Figure 10.3 A pod of seven weanlings sleeping in the dunes several meters away from the harem where they were born.

average weanlings, that is, up to 225 kg or more. Moreover, these milk thieves get additional essential proteins.

There is a significant sex difference in this behavior. Male weanlings are much more likely to reenter the harem to attempt to steal milk than female weanlings, and they are more persistent in continuing their intrusions after being rebuffed. While both sexes run the same risk of being injured, males increase their size, which benefits them in later life. Increased mass at weaning is positively correlated with size later in life, which is associated positively with dominance status and increased probability of mating. The gain in size for females is not proportional to the risks incurred.

Besides milk-stealing, there are other physical sex differences indicating that the pattern of development serves and is shaped by reproductive competition. Males outweigh females at birth and at weaning (Le Boeuf et al. 1989). Males molt the black natal pelage about a week later than females, which may help them mimic young suckling pups and give them an advantage in stealing milk (Figure 10.4). In addition, canine tooth eruption is later in males than females, which may serve the same purpose (Figure 10.5). Males are weaned one full day later than females, and the extremely large weaned pups are always males. These sex differences show that the male strategy to get an edge in sexual competition begins early in life.

Newly weaned pups weigh 132 kg, on average, 59% of which is blubber (Rea and Costa 1992; Kretzmann et al. 1993). Weanlings live off this fat store while fasting from food and water for about 2½ months, while they learn to swim and dive in the shallows of the rookery before going to sea on their maiden foraging trip. From the start, opportunities for weanlings depend, in part, on the time of birth and weaning (Reiter et al. 1978). Consider the different environments of pups born and weaned early in the breeding season, during the middle of the breeding season, and late in the breeding season (Figure 10.6). The early born have more time to learn to swim and

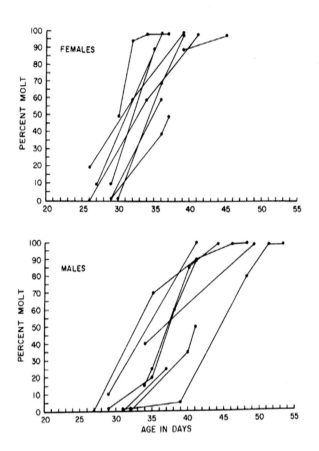

Figure 10.4 Sex difference in molting; males molt later than females.

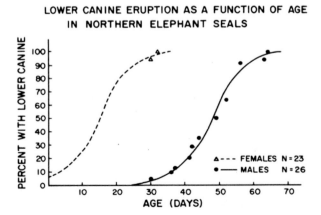

Figure 10.5 Lower canine tooth eruption of pups varies with sex; tooth eruption is later in males than females.

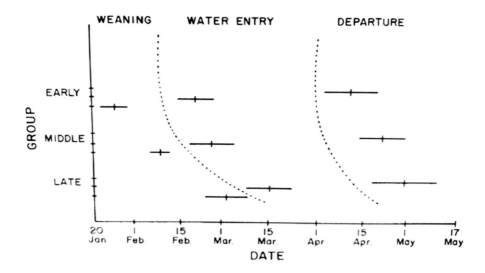

Figure 10.6 The timing of births during the breeding season presents more opportunities to the early born for stealing milk and learning to swim and dive that benefits both the pup and its mother.

dive (74 days) than pups born late (56 days), and they have more opportunities to steal milk whereas the late-born do not because few nursing females remain in the harem at this time. Pups born early in the breeding season enjoy the highest parity rates as adults (16%), compared to 9% for those born in the middle of the breeding season and 8% for the last-born. Of course, the mothers also benefit from the timing of births. Prime or old females give birth early in the breeding season; the youngest females give birth late.

Within two weeks after weaning, weanlings encounter water for the first time in the form of standing freshwater ponds, tide pools, or shallow water in protected coves or beaches. They approach water with great curiosity, initially nosing or smelling it as they timidly immerse their bellies in it. Initial attempts at swimming and diving are awkward and uncoordinated, involving much flapping around, but improvement is rapid. Soon, they gravitate to the water at dawn and dusk and return to the dunes to spend the rest of the day or night sleeping. With each successive day the weaner becomes more accustomed to the new medium and spends increasingly more time in the water.

After about two weeks, weaners venture out beyond where they are touching bottom and begin to make short dives lasting several seconds. After three weeks of practice, they can dive for several minutes and sleep underwater for up to seven minutes, as well as hold their breath for up to eight minutes while sleeping on land (sleep apnea). They begin to swim along the shore away from the point of water entry. Shortly before they leave the rookery on their first foraging trip, they are spending about an equal amount of time in the water as on land, most of it at night. They do not feed while learning to swim and dive. The mean duration of dives increases from 1.9 minutes at initial water entry to 6.1 minutes just before they leave the rookery on their first trip to sea. Throughout this period near the rookery, they are in water less than 12 m deep.

The pup's move from land to sea is especially abrupt and demanding, requiring extreme adaptations for breath-holding and withstanding high pressure, while finding food (Reiter et al. 1978; Thorson and Le Boeuf 1994; Le Boeuf et al. 1996). As mass decreases during the post-weaning fast, hematocrit, mass specific blood volume, and myoglobin concentration increase significantly. Mass-specific oxygen stores increase by 47% over the post-weaning fast, or 69% from the suckling period to the time when the seals go to sea.

The weanlings leave the rookery in early to late April when they are about 3½ months old. Their departure coincides with coastal upwelling in the area that is positively associated with enrichment at the base of the food chain, which has positive effects on fisheries. It is not clear, however, that this helps them find food. At departure, weanlings weigh about 99 kg on average, having lost 25%–30% of their weaning mass. They are now accomplished swimmers, but all of their diving experience has been in shallow waters.

The weaner swims along the bottom and rises to the surface to breathe as it crosses the continental shelf until it reaches the continental slope. Here, it faces the deep water of the open ocean, and an entirely new world. It is alone, naïve, and knows not what to eat, where to find its prey, or how to catch it and eat it. The weaner doesn't know its predators and how to avoid them. It has received no training in how to survive at sea from its mother and certainly not from its father. Surviving the first trip to sea is perilous; 50% or more of the weanlings in an average year, will not make it. The second and third trips to sea are marginally better. High mortality at sea during these early years appears to be due mainly to insufficient foraging, predation, or accidental contact with the environment, fishing or ghost nets, or ships strikes.

Finding, catching, and consuming sufficient prey may be the greatest problem. Patricia Morris (pers. comm.) showed that during the first two foraging trips, a 15- to 16-month-old yearling weighs about 100 kg, the same as it weighed when it began its first foraging trip at 3½ months of age. The only noticeable difference is that the lean-to-fat ratio of the yearling increased over this period and so did its standard length by 10 cm. Moreover, during the first trip to sea, heavy pups gained little or no weight while some small pups doubled their departure mass.

Young seals on their first few foraging trips may get ensnared in fishing nets, ghost nets, or assorted debris that is increasingly prevalent in ocean waters (Hanni and Pyle 2000; Campagna et al. 2007; Jepsen and Nico de Bruyn 2019). The degree to which this is a significant cause of mortality is unknown. White sharks, *Carcharadon carcharias,* and killer whales, *Orcinus orca,* take an unknown quantity of the weanlings (Ainley et al. 1981; Le Boeuf et al. 1982b; Klimley et al. 2001; Le Boeuf 2004). Why not? The weanlings are fat burgers, appealing nutritious morsels for a predator that wants a high-energy snack. Avoiding predators is not enough, however; the young seals must find, catch, and consume sufficient prey. This may be difficult as reflected by the lack of mass gain, as we shall see. Another obstacle is that the young seals going to sea for the first time are positively buoyant and require more effort to descend to depth to reach their prey than adults.

The prey of northern elephant seals during the first trip to sea has not yet been determined. The diet of southern elephant seals during the first trip to sea is predominantly krill (*Euphausia* species)(Walter et al. 2014; Orgeret et al. 2018).

In any case, when weanlings first encounter deep water beyond the continental shelf on their first trip to sea, their average dives last 10 minutes and the maximum dive duration is 22.3 minutes; their dive depth averages 200 m, with a maximum depth of 553 m. There is no gradual transition from brief and shallow dives near the rookery to diving long and deep in the open ocean. They adjust almost immediately to diving deep and long. Like adults, the majority of their time at sea is spent underwater. In brief, the diving pattern of these young seals approaches that of adults on the first trip to sea!

Perhaps no other marine mammal makes such a drastic transition in such a short time. The transition is in behavior as well as physiology. Concomitant with improvements in swimming and diving performance are significant increases in oxygen storage capacity and decreases in diving metabolic rate. The more the young seals dive, the faster the metabolic rate declines, the longer the dive durations, and the more the time spent submerged.

Being a capable diver, however, is not enough to survive; only 25% of the female weanlings that go to sea live to age 3 or 4, when females give birth for the first time (Le Boeuf et al. 2019). During the years 1971–1988, the mortality of a sample of over 8,000 elephant seals born was 63% to year 1, 74% to year 2, 81% to year 3, and 84% to year 4. Survival to breeding age is lower among males than females, in large part because they take longer to mature and are more likely to die before breeding for the first time. Only about 5% of male weanlings survive to age 8, the time when males begin to have an opportunity to mate.

The developmental period we address here includes the first three years of life. This cut-off is imposed because the land–sea schedule of males and females diverges at age 3 because some females give birth for the first time at this age; that is, giving birth defines them as adults although they continue to increase in size (Reiter and Le Boeuf 1991). Males continue to develop and grow but do not attain physiological sexual maturity until five or six years of age, which is still a few years from reaching social maturity.

Juvenile foraging trips 1, 3, and 5 occur over summer from April to October and last 123 to 173 days. Foraging trips 2 and 4 are made over winter from November to April and last about 150 days. The age in months for juveniles at each consecutive foraging trip is 3–9, 10–15, 16–21, 22–27, and 28–33, respectively.

The initial foraging trip. The first trip to sea is the most challenging. A study of 8,362 weanlings tagged in the years 1971–1978 determined that on average only 37% survived the first trip to sea. Survival varied over the years from 20% to 49%. Survivorship was lowest in 1983, a severe El Niño year. Although the migratory path of yearlings on their first trip to sea has not been determined, recovery of tagged seals along the coast and observations at sea suggest a general northward movement

(Condit and Le Boeuf 1984) like that of older juveniles and adults (Le Boeuf et al. 1993). Yearlings born on California rookeries have been observed off the coasts of northern California, Oregon, Washington, and British Columbia. Moreover, more deviant sightings are reported in this age group than in any other, which suggests that some young seals get lost. For example, a yearling was sighted on Amaknak Island in the Aleutian Island chain, 4,000 km from its birthplace on Año Nuevo (R. Nelson, pers. comm.), a yearling was seen on Midway Island, 4,700 km west of its birthplace in southern California (G. Blazs, pers. comm.), three juveniles landed in Japan, approximately 8,000 km from the west coast of North America (Kiyota et al. 1992; Y. Naito, pers. comm.), and over the last decade there have been several sightings of yearlings and juveniles in the Sea of Cortez in Baja California, Mexico, 1,655 km from the nearest rookery (Condit and Le Boeuf 1984; Aurioles et al. 1993; S. Mesnick, pers. comm.). Yearlings returning from their first trip to sea weigh less (90 kg) than they did when they set forth (96 kg). The fat in their bodies decreases by 41% (Kretzmann et al. 1993). The only noticeable difference is that their lean muscle-to-fat ratio increases and they are about 10 cm longer. Low survivorship and failure to increase their mass during the first trip to sea suggest that the transition to pelagic life and feeding is difficult.

Nevertheless, dive records of the initial days at sea reveal a pattern similar to that of adults (Thorson and Le Boeuf 1994). The yearlings dive continuously and are submerged for about 85% of the time. Mean dive duration is about half that of adults, with a maximum of 22.3 minutes. Surface intervals are brief, between 1.4 minutes and 1.8 minutes. These naive seals begin to dive deeply as soon as they reach deep water. Diving instruments indicate a mean depth of 200 m and a maximum depth of about 550 m. Total oxygen stores increase by 14.2% over the five-month first trip to sea, culminating in oxygen stores that are 84% of those of adult females.

Does weaning mass predict survival? Although the mass of weanlings varies greatly, with animals in the highest weight category being almost twice as large as those in the lowest weight category, there is no significant relationship between mass at weaning and survival to age 1 or to age 2 (Le Boeuf et al. 1994). The fattest are not the fittest.

Juveniles 9–17 months of age. Survivorship over the ensuing three migrations, when juveniles reach two years of age, increases annually. For example, whereas 37% of the sample of 8,362 weanlings survived to age 1, 26% survived to age 2, 19% to age 3, and 16% to age 4. That is, survivorship rate improves with experience at sea and age. Foraging skills improve by the second trip to sea, as evidenced by a 60% increase in return mass over departure mass. With each successive trip, the seals become more proficient at putting on mass. Mean mass gain of females during the juvenile years is on average 0.51 kg per day, half that of adult females. Juvenile males put on significantly more mass, 0.68 kg per day, than juvenile females.

The migratory paths of juveniles are northerly to the latitude of Washington State and then they diverge over a wide area. Distances traveled increase with age and experience. On the second trip to sea, the farthest northerly point reached is latitude 48°N compared with 50°N on the third trip. Similarly, maximum distances traveled are greater for older juveniles, in some cases paralleling the distances traveled by adults (Le Boeuf et al. 1993). Migratory routes and distances traveled by 1.8-year-old juveniles are as far as southern Alaska. Horizontal swim speed increases with age and experience, from 56 km per day on trip 2 to 68 km per day on trip 3, and to 85 km per day on trip 4. One juvenile traveled 116 km in one day! Juvenile males traveled slightly faster than juvenile females, and both sexes traveled more slowly than adults. In general, juvenile males remained closer to the continental margin than juvenile females, a sex difference observed in adults (Le Boeuf et al. 1993, 2000a), and males displayed more benthic dives than females and females favored pelagic foraging dives, which is the pattern in adults.

Over the course of three trips to sea, diving performance approaches the level of adults: dive rates decrease, and dive duration and dive depth increase. By the end of the fourth trip to sea, at two years of age, average dive durations and dive depths are on a level with those of adults. Although only two-thirds the size of young adult females, two-year-olds are seasoned, accomplished divers.

These studies suggest the following conclusions regarding the development of diving in this species.

(1) The greatest increase in breath-holding and oxygen storage capacity occurs during the post-weaning fast, when the weanlings are learning to swim and dive near the rookery.

(2) Diving experience gained near the rookery is more closely related to dive duration requirements than dive depth requirements at sea. The mean duration of dives near the rookery is about 60% of the mean duration of dives once the yearling goes to sea. In contrast, average dive depths during the practice period, being depth limited, are only a fraction (3%) of mean dive depths exhibited by yearlings once they reach the open ocean.

(3) The transition to a pelagic existence is difficult. The probability of dying during the first trip to sea is higher than at any other time in life, and those that survive do not gain weight during the five months of foraging.

(4) Improvement of diving skills continues to two years of age. No period of acclimation seems necessary for achieving modal diving performance during each of the first four migrations.

(5) Juvenile migrations finish and refine the development of diving to adulthood. By the end of the fourth trip to sea, when juveniles are approximately 27 months old, the diving pattern is essentially the same as that of adults.

(6) Sex differences in the foraging pattern and foraging location, similar to those seen in adults, begin by the third or fourth trip to sea. As with adults, there is a tendency for juvenile males to remain closer to continental margins than juvenile females and for females to prefer open waters. Similarly, juvenile males exhibit a high frequency of flat-bottomed benthic dives and juvenile females favor pelagic foraging dives. This suggests sex differences in prey consumed.

Although juvenile seals approach the diving capacities of adults by age 2, there are two elements in the diving pattern where they differ from adults. Juveniles exhibit twice the percentage of drift dives and twice the percentage of extended surface intervals as adults. These two events are closely associated with foraging in diving records. The presumed function of drifting is to allocate oxygen reserves to internal physiological processes, such as the metabolic cost of processing food, at the expense of reduced locomotion or sleep or rest (Crocker et al. 1997). Extended surface intervals are associated with long bouts of foraging dives that suggest gorge feeding. Dive records are consistent with the hypothesis that when the seal cannot process all of the prey it has consumed while drift diving, it continues the process at the surface. These differences suggest that juveniles are not as efficient as adults in obtaining and processing food.

11   Sleep When You Can

The more you sleep, the less you live.

Polish saying

Sleep in animals is a behavioral and physiological state characterized by reduced responsiveness to external stimuli, minimal movement, and homeostatic regulation, and the pattern is usually repeated on a 24-hour basis. All mammals and most animals need to sleep. The sleep patterns vary widely among species; some sleep at night, others during the day, and others partition sleep into multiple bouts during the 24-hour day.

All animals must recover from lost sleep, and they suffer serious consequences and may die from sleep deprivation. Deep sleep follows sleep deprivation. Large animals sleep less long than smaller animals.

When do elephant seals sleep in both environments they occupy? On land, observations suggest that the answer is whenever they can (Box 11.1). They exhibit polyphasic sleep, short bouts of sleep spread out over the 24-hour day. During the breeding season, the alpha male and other high-ranking males have the luxury of sleep only when most other males are asleep. Otherwise, they are busy keeping other males away from the females in the harem. The alpha male must remain awake to prevent others from mating. Those that sleep lose in reproductive warfare (Figure 11.1). The best opportunity for a long snooze is on a hot sunny day when all residents are too listless to be active and remain awake. All activity in or around the harem comes to a standstill. Otherwise, the circumstances only allow high-ranking males to steal a few "catnaps" between chasing others, being chased, or attempting to mate. It is almost as if the alpha male sleeps with one eye open; he naps lightly but is alert to the slightest disturbance, such as the sound of a female being accosted by a suitor. Getting sleep is a challenge for the alpha male because the harem is usually like a high-activity circus; there are incessant vocal challenges between males, the galumphing of two-ton seals rushing to thrash a competitor or "running" from an enraged bull through females and pups to save their skin, the protest calls of females that don't want to mate, and the cries of pups that want to be fed. Mating trumps sleep, so high-ranking males must manage with little sleep. It is likely that these males suffer from sleep deprivation after three months of competing to mate. Indeed, when the breeding season ends and the last female has departed and gone to sea, the high-ranking males spend about two

---

Box 11.1 What Do They Do at Night?

Elephant seals do the same things at night that they do during the day. This was revealed from a camera housed in a blind that photo-multiplied available light so that on a moonlit night the video screen revealed exactly what was going on. Males threaten and fight each other. They mate or attempt to mate. The alpha male prevents others from mating. Females nurse their pups and squabble with each other to protect their pups. They vocally protest being mounted when they are not interested and even when they are interested in mating. Indeed, the seals may be more active at night than during the day. They are least active and most likely to sleep on warm, sunny days. The difference between night and day is most obvious from the behavior of the birds – several species of sea birds and shorebird species such as western gulls, Heermann's gulls, turnstones, and willets – that are not flying around at night but rather are still, resting in groups and usually facing in the same direction.

Figure 11.1 An adult male in deep sleep away from the competition in harems. (A black and white version of this figure will appear in some formats. For the color version, please refer to the plate section.)

solid weeks sleeping on the beach before they go to sea. Evidently, they require sleep more desperately than food despite having fasted for over 100 days and having lost about 40% of their arrival weight.

Adult females have a similar problem, but the circumstances are different. The enemy of sleep for females is males attempting to mate with them at any time and regardless of whether they are capable, recalcitrant, or willing. It is hopeless for a lone female removed from the harem; she is accosted continuously by numerous males even when she is pregnant and clearly not in estrus. In a harem she is less vulnerable to male disturbance. Although she may be accosted by the alpha male at any time, he has many other females to pursue, and he protects her from the mating attempts of other males in the area. No female is free from the mating attempts of males; males "test" pregnant and nursing females repeatedly, paying no attention to their interests.

Females on the periphery of harems are most frequently disturbed by males because they are more accessible to the numerous peripheral males surrounding harems than females in the center of the harem who are close to the alpha male. After giving birth, a female has another reason not to doze off. She must see to the safety of her pup and this amounts to nursing it, keeping it near, preventing it from wandering, preventing neighboring females from stealing or biting it, and reuniting with it quickly after momentary separation due to male disturbance or waves crashing into the harem. A female sleeps lightly when her pup is sleeping quietly next to her. She is ever attentive to her pup's call or its nudge against her nipples when the pup wants to suckle. Her catnaps seldom last long. By and large, females sleep when their pups are near and safe, and they are not disturbed by other females or males.

Elephant seals do not show a preference for sleeping at night or during the day. We confirmed this during the 1970s. We employed a large video camera, a photomultiplier, that allowed us to see the seals at night with only the barest illumination from the moon. The behavior of both males and females in harems was indistinguishable from their behavior during the day. The most obvious change from day to night was a change in the behavior of the Western gulls and other marine birds on the beach.

Newly weaned pups, as well as young juveniles, spend a great deal of the day sleeping alone or in groups removed from the harems. They sleep for long periods because they are usually undisturbed. As weaned pups age and learn to swim and dive, they spend increasingly more time in the shallow waters at night and then sleep long and soundly in the sand dunes during the heat of the day.

During the molt, which lasts about a month, all of the seals onshore spend the majority of their time sleeping. Males are not aggressive to each other (the gonads have regressed in size and the production of sex hormones has abated) and sleep side by side. The same is true for adult females who molt at a different time of the year than males. Again, there is no clear diel pattern of sleep; the seals are neither nocturnal nor diurnal but rather sleep opportunistically at any time.

What about sleep at sea? Since elephant seals spend most of the year at sea, and 90% of this time they are submerged, do they sleep on the surface or underwater? These questions have fascinated us since we got our first diving record (Le Boeuf et al. 1986) showing that virtually continuous diving was normal for most of the time the seals are at sea.

The sleep pattern at sea mimics the sleep pattern on land or vice versa. Elephant seals dive virtually continuously day and night during long periods at sea lasting from two to eight months (Chapter 10). On average, each dive lasts about 20 minutes and is followed by about 2–3 minutes breathing at the surface. This cycle is almost continuous during the day and the night. If the seal slept at the surface between dives, it would be sleeping only 2.4 hours per day. This is not much time compared to the amount of sleep observed in other mammals, including humans, which is about 8 hours or 33% of a 24-hour day.

Moreover, observations of seals sleeping on land (Bartholomew 1954; Kenney 1979; Huntley and Costa 1983; Blackwell and Le Boeuf 1993; Castellini et al. 1994) suggest that elephant seals at sea sleep while holding their breath, that is, while submerged, and they may or may not continue sleeping when they come to the surface to breath. Seals sleeping on land exhibit sleep apnea; they hold their breath for about 20 minutes (duration depends on the sex, age, and size of the animal) and then breathe for 2–5 minutes and then hold their breath again, repeating this pattern over and over for hours until they wake up or are disturbed. Watch an elephant seal sleeping in the dunes at the Año Nuevo mainland. You might see a young male with its head submerged in a tidepool (Figure 11.2). You might think it is dead because it is not moving but if you look at its chest you can see the heart thumping and calculate its heart rate, which slows down during apnea. After about 20 minutes or so, it will lift its head out of the water, breathe slowly and perhaps yawn, and in a few minutes put its head underwater again and sleep while holding its breath. The resemblance to the diving pattern at sea is so obvious that we call this sleep pattern "terrestrial dives."

Studies of animals sleeping in the laboratory, with instruments attached to monitor sleep, show that sleep always occurs during breath-hold (apnea) and may continue or not during breathing (eupnea). Thus, sleep on land suggests that sleep at sea occurs mainly while the seals are submerged. When might this be? Sleep is most likely while drifting, or possibly during the descent or ascent phase of dives. Drift dives, described in Chapter 9, are characterized by stroking descent and gliding to a depth of about 200–400 m followed by a segment of passive drifting, which takes up about half of the total dive duration, ending with ascent to the surface. These dives are associated with long bouts of feeding and may facilitate digestion and, at the same time, permit sleep and rest (Crocker et al. 1997). The seals do not feed while drifting and they do not stroke; they float up slowly when positively buoyant and flow down when negatively buoyant. A similar case is made for sperm whales sleeping while drifting near the surface with the head up or down (Miller et al. 2008). The whales are inattentive while drifting and can be approached and touched. The problem for elephant seals, though, is that there are few drift dives in the 24-hour day. On average, females spend about

Figure 11.2 A subadult male sleeping in a tidepool with his head submerged in water.

48 minutes per day drifting (Crocker et al. 1997). Even if the seals spent all of this time sleeping, which is not clear, only 3.3% of the day would be spent sleeping.

Might elephant seals might sleep during the descent or ascent portion of dives? That is, might they dive up or down on "automatic pilot" with reduced vigilance? The reason for this question is that the working part of dives is the bottom part of dives. This is when they chase, catch, and consume prey. The argument is that the only the time a seal must be awake is at the bottom of dives. From another perspective, the time underwater for seals might serve the same function as sleep because heart rate and metabolism are lowered. That is, the seal may not have to sleep as much as other animals because the physiological changes associated with diving provide similar benefits as sleep. More study is needed to test these speculations.

Dolphins (Mukhametov et al. 1977; Ridgeway 2002; Lyamin et al. 2007), some whales (Lyamin et al. 2002), and four species of otariids – notably Alaska fur seals (Lyamin et al. 2018; Kendal-Barr et al. 2019) – employ unihemispheric sleep while submerged or diving. One hemisphere remains fully functional, while the other goes to sleep. The hemisphere that is asleep alternates, so that both hemispheres can be fully rested. These animals may also open one eye and close the other. This allows the animals to sleep part of the brain while moving through the water and being vigilant for predators. For example, dolphins schooling in a circular pattern sleep with the outside eye open scanning for predators while the inside eye is closed. Many other animals and birds use the trick of allowing one side of the brain to sleep while the other side is awake and "attentive."

Northern fur seals switch from bihemispheric sleep on land to unihemispheric sleep at sea (Lyamin et al. 2018). Bihemispheric sleep is typical of phocid seals on land as well as at sea. Unihemispheric sleep has not been observed in elephant seals or other phocids. Perhaps deep diving makes this unnecessary. The walrus is like elephant seals (Lyamin et al. 2012), utilizing bihemispheric sleep on the surface (ice) as well as in the water.

It would be neat to have an EEG cap that can be fitted to the head of a free-ranging seal to determine its sleep pattern during diving. This is not so far-fetched; this innovation is well on its way to being launched (T. Williams and J. Kendal-Jackson, pers. comm.). I'm eager to see the results.

12   What Is All the Noise About?

We like no noise unless we make it ourselves.

Marie de Rabutin-Chantal, Marquise de Sevigne

Imagine 300 mothers with newborns crowded close together on a beach. The mothers want space to nurse and protect their pups; the pups want to be close to those nipples that satisfy hunger. A pup cries out for its mother to roll on her side so it can suckle. If she doesn't respond, the pup may shriek and nip its mother's belly in frustration or wander off and seek milk elsewhere. The mother emits her warbling call to her pup telling it where she is, but a neighbor nearby approaches to check out the pup to see if it is hers. The mother directs a loud threat vocalization to the neighbor approaching her pup, as well as to all other neighboring females warning them to keep their distance. Hundreds of interactions such as this go on all the time and provide a continuous cacophony of high-intensity background noise. The din is intensified by males competing to mate with females, whether they are willing or not, threatening each other with loud clap-threats and chasing them away, or fighting fiercely in the middle of the harem, and upsetting the spacing between nursing mothers. This adds to the cries, threats, and clamor among neighboring females and their pups. In addition, there is the constant loud pounding of the surf that forces females on the periphery to crowd closer to females near the harem center. The gulls shriek overhead competing to feast on discarded placentas. The sea lions bark in the background. And the wind and the rain come and go. An elephant seal rookery during the breeding season is a noisy place (Bartholomew and Collias 1962).

Elephant seals are a conspicuously vocal species and vocalizations are produced by individuals of both sexes and all ages. Each vocalization communicates a message and is directed to a specific receiver. I address the general purpose of the principal vocalizations that one hears during the breeding season. The physical properties of the vocalizations are another matter and are not appropriate here.

Adult males. All of this noise puts pressure on individuals to "speak" loudly to communicate their intentions. This is especially true for a male threatening another male. The message signals incipient attack: "I am Thor, go away or I'll bite, batter, and thrash you to shreds." The message has to rise above the background noise to be effective. The threat vocalization – also called clap-threat or belch-roar – is loud, resonant, metallic-sounding, and pulsed; it has excellent carrying power, making it easy for the receiver to hear it from about 25 m away over the background noise and localize where it is coming from. It is always produced from the same stereotyped posture, with the forequarters elevated to the maximum, braced by the extended fore flippers, and the long proboscis extended down into the fully open mouth. The proboscis has nothing to do with the sound emitted. In this position, the male is best positioned to attack or meet the attack of his opponent, or back away and flee while facing the victor. Threat vocalizations of males are the most prominent sounds of these seals. If you visit a rookery during the breeding season, you will hear males bellowing well before you see them.

A recent study shows that threat vocalizations of male elephant seals are among the loudest calls measured in terrestrial mammals, exceeding those of African elephants and howler monkeys and comparable in level to those of lions and North American bison (Southall et al. 2019). Males do not vary the loudness of their threat calls; they are stereotyped and consistent in loudness. The sound level also does not vary according to the individual's dominance status. Indeed, I recall the threat calls of one alpha male as being relatively quiet, more like a loud click, compared to other males, which means that other males had to listen carefully or they paid a price. The importance of the threat vocalization is that it communicates the identity of a specific individual. Males in competition around a harem must recognize the calls of the males that dominate them and those that they dominate. Observations over the years indicate that they develop a good memory of competitors and their calls and act appropriately. Mistakes occur occasionally, and the perpetrator that threatens another male suddenly realizes that he made a mistake and the other male is dominant to him, not a subordinate. The perpetrator slinks off, and, to my eye, looks "embarrassed."

Imagine a rookery where all the threat vocalizations of males are similar but distinctly different from the threat vocalizations of males from other rookeries. This is what we discovered in 1968 and 1969 (Le Boeuf and Peterson 1969a; Le Boeuf and Petrinovich 1974a, 1974b, 1975). The feature of the vocalization that was similar across males was the rate at which the vocal bursts in the threat vocalization were delivered, or what we called their pulse rate (Figure 12.1). Most males deliver four to eight vocal bursts in a vocal bout. For example, the pulse rate of males at Año Nuevo was 1 pulse per second. Honk, honk, honk, honk, honk. This was not so unusual; we simply thought that this was how males "talked." But what caught our attention was that the threat vocalizations of males at other rookeries were different. Males at Isla de Guadalupe and San Miguel had faster pulse rates of 1.8 pulses per second. The pulse rate at San Nicolas Island was the fastest, 2.5 pulses per second. The differences between the rookeries were obvious to our ears; we didn't need sophisticated measuring instruments or statistical tests to convince us. Of course, we made tape recordings of the threat vocalizations of males at each one of these colonies. These geographical differences in vocal behavior resembled vocal dialects in birds and humans. Local dialects are consistent differences in the predominant song or call of adults from different populations of the same species. We knew about dialects in birds and humans, but they were unknown in other mammals at the time. How did they arise? Did they change over time? What was their function?

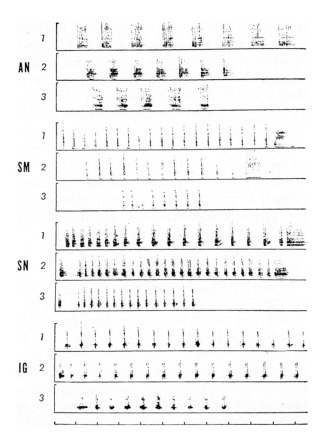

Figure 12.1 Dialects appear in the threat vocalizations of males recorded at four different colonies in 1968 and 1969. The calls of three males from four colonies are shown: AN – Año Nuevo Island; M – San Miguel Island; SN – San Nicolas Island; and IG – Isla de Guadalupe. The pulse rates (pulses per second) shown on the baseline are consistent at each colony but vary across colonies; AN is slow, SN is fast, and SM and IG are intermediate. From Le Boeuf and Peterson (1969a).

We found that the mean pulse rate at Año Nuevo Island increased every year from 1968 to 1972, up to 1.47 pulses per second, but still remained slower than the mean pulse rate of southern rookeries. Since 43% of the breeding males at Año Nuevo were immigrants from southern rookeries with faster pulse rates, we reasoned that the dialect drift resulted from the faster talking males migrating to Año Nuevo causing the elevation in pulse rate. In 1974, the pulse rate of males from San Nicolas had done the opposite; it had decreased to 1.5 pulses per second. Moreover, the pulse rate of males at San Miguel Island did not change significantly from 1969 to 1972.

We left the dialect study hanging and unresolved because we had other pressing research matters to address at the time. Fifty years later (this is not a typo), the original study was repeated to check on the status of the dialects (Casey et al. 2018). Male calls were sampled at the same four rookeries in 2014 and 2015 that were sampled in the original study in 1968 and 1969.

The dialects had disappeared completely. Why? The best explanation for the appearance and disappearance of these dialects is that they reflected the immigration pattern of the seals as the population continued to expand and recover from near extinction. In brief, consider that after the nadir of the population in the late 1890s, as the population began to recover, the seals first immigrated from Guadalupe to southern California. As the colony at San Miguel grew, seals immigrated to proximate rookeries like San Nicolas Island and north to Año Nuevo (Figure 2.2). It is likely that the immigrants of this time had a wide range of pulse rates, but they averaged about 1.8 pulses per second. We reasoned that the males that settled at San Nicolas Island were originally "fast-talkers." The males that settled at Año Nuevo were originally "slow-talkers." The founders set the tone for the original appearance of dialects, but with continued immigration from mother colonies like San Miguel Island with a pulse rate of 1.8 pulses per second, the average pulse rate of males at Año Nuevo got faster and those at San Nicolas got slower. With time and continued immigration, the dialects were completely lost. Statisticians call this regression to the mean. This brought us back to considering the importance of dialects. Whereas ornithologists explained bird dialects as keeping populations isolated and preventing gene flow (Nottebohm 1969, 1972; Thielcke 1969), it was just the opposite in elephant seals. The dialects were a result of isolation, which, in time, were lost with continued immigration. Dialects in elephant seals were a moment in time, a by-product of population expansion and immigration.

We can illustrate this point with a human analogy. Imagine the speech of people from Alabama and Massachusetts years ago. They both spoke English, but you could easily differentiate one from the other because of their different dialects. Now imagine that immigrants from a small town in Alabama begin emigrating and relocating to a small town in Massachusetts. With continued immigration, and depending on the rate, the speech of immigrants from Alabama would start to take on elements of the speech of people in Massachusetts, and vice versa. With continued immigration from the south to the north, the differences in vocal behavior of the immigrants and natives in the small town in Massachusetts would disappear.

There was another feature of the recently recorded threat vocalizations of male elephant seals that changed (Casey et al. 2018). They got more complicated over time, compared to the simplicity of the early vocalizations. There was more variability in the calls of individuals at each colony in the recent sampling than the original sampling. We think increasing signal complexity – such as a trill at the beginning or end of the string of bursts, or syncopation in the string of bursts – was a response to increasing colony size. Over the 50 years when samples were obtained, a 10-fold increase in the population size of each colony, except Isla de Guadalupe, occurred (Le Boeuf et al. 2011). As the number of males increased, structural differences in calls become increasingly useful to individuals in the recognition process, or they got hurt.

Subadult males. Young males, approximately five to seven years old, make similar threat vocalizations as adult males but they are not prominent near harems. The reason is that a subadult male vocalizing gives away his position, and if he has snuck into a harem or is close to it, dominant males will hear him and chase him out or approach without warning and bite him viciously on the back. In this situation, the subadult will emit a high-pitched whimper while backing away trying to prevent being mauled or bitten.

Subadult males can be very vocal in the dunes away from harems while fighting for dominance with each other. The calls have a higher frequency than those of adult males. Some subadults will practice their calls in coves surrounded by high rock walls. The sounds are amplified, which seems to provide positive feedback to the callers. Practicing males have kept me awake many nights because our sleeping quarters on the island were adjacent to one such "rehearsal" cove.

Adult females. Adult females with pups make two primary vocalizations. One is a loud, harsh, deep, rasping, and grating roar that signifies threat or discontent (Le Boeuf et al. 1972; Cox and Le Boeuf 1977; Christenson and Le Boeuf 1978). The threat is heard in two different situations. It is used in altercations with neighboring females over space, in defending her pup against attacks by neighbors, against a bull resting and holding down her pup, or against any other threat, such as a human approaching her and her pup. The roar is long and lasts three to five seconds. The threat is delivered with the mouth open and directed at the source of the threat (Figure 12.2) The threat vocalization is even more prominent when directed at males to reject mating attempts. Here, the female is voicing her discontent. She protests

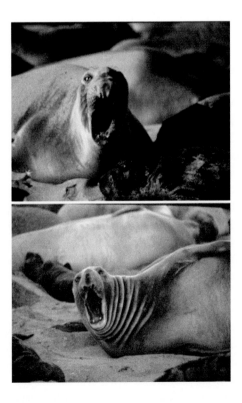

Figure 12.2 Adult females with pups issuing threat vocalizations.

vocally and rejects or tests males that attempt to mount her (see Chapter 8). One gets the impression that the female is giving a firm "no" to males. If one were to design a vocalization that sounds like a firm "no," this is it. Females routinely reject most mating attempts this way. One exception is when a female is mounted by the alpha male; after a while she may stop vocalizing and accept him. We argue in Chapter 8 that it benefits a female to recognize that the suitor is the alpha male, the male best suited to sire her offspring, and best for her and her offspring. Of course, females stop rejecting male mounts at the peak of estrus. Also, she does not threaten males waiting to intercept her as she returns to sea to forage (Le Boeuf and Mesnick 1991). In this situation, she is proactively receptive and invites mating attempts, which we interpret as buying safe conduct in a dangerous situation.

The other principal vocalization of mothers is the pup attraction call. As soon as she gives birth, the mother wheels around to face the pup and utters a high-pitched yodel directly in the pup's face, while jerking her head up and down. She may also smell or taste the pup, using multiple senses to ensure identification. The pup responds with a vocalization of its own. These duets may continue for several minutes and is important in establishing the bond between mother and pup, which they use to remain in close contact. Later on, if the pup strays or is separated for any reason from its mother, the pups calls and the mother answers or vice versa, and ideally they get together again. Mothers recognize the vocalizations of their pups and will move toward them. Less reliably, they will also move to other pups vocalizing to see if this is her pup, even if their own pup is at their side. Some mothers will bite pups that are not their own. Consequently, mothers are wary of neighbors and fear and threaten them to keep their distance.

Pups. Pups vocalize under a number of situations: (1) when separated from their mother, (2) when restrained from moving because they are pinned down by a bull or a neighboring female, (3) when bitten by a neighboring female or run over by a charging bull, and (4) when hungry and unable to suckle because of the mother's position. In all of these circumstances the pup is distressed. It raises the forequarters, elevates the head, and opens the mouth, making a conspicuous display of the bright pink buccal mucosa. Mothers respond to pup calls; bulls are oblivious to them even when a pup is pinned beneath them (Figure 1.1).

Orphaned pups vocalize a lot and this either gets them an opportunity to suckle from a willing female, often a female that lost her own pup, or it attracts the attention of mothers that bite and shake them to prevent them from stealing milk intended for their own pups. After pups are weaned, some cry forlornly through the night for the mothers that abandoned them. Or, perhaps, they do this to keep us humans from sleeping.

The sheer volume and variation of sounds made by elephant seals demonstrate that vocal behavior plays an important role in their social life. It helps to know what the vocalizations that you hear mean if you want to understand the behavior of elephant seals on a rookery during the breeding season.

Reception of sounds. One might expect elephant seals to utilize the auditory sense in the ocean because transmission of sound is enhanced in water, while the visual sense, in contrast, is hampered in water by poor conduction of light. As far as we know, elephant seals do not vocalize while submerged. They are, however, excellent receivers of acoustic signals. They possess superior low-frequency hearing capability on land as well as the ability to receive low-frequency sounds emitted from great distances in the ocean (Costa et al. 1996, 2003; Kastak and Schusterman 1998). Since elephant seals dive routinely to depths of 600–1,000 m, this puts them in the deep sound channel, a horizontal layer of water in the ocean where the speed of sound is at its minimum. The channel acts as a waveguide for sound, and low-frequency sound waves within the channel may travel thousands of miles before dissipating. Only a few other mammals, sperm and beaked whales, dive to these depths (see Chapter 13).

Being able to receive low-frequency sounds at depth from remote sources might make elephant seals vulnerable to disturbance from sources such as air gun shots for geophysical surveys, transmissions by the US Navy, oil tankers, and various other sources of noise. One of these acoustic sources that received a lot of attention, because some thought it threatened marine mammals, was the Acoustic Thermometry of Ocean Climate (ATOC) experiment (Munk et al. 1995). Its aim was to determine the rate at which the ocean was warming. The objective was to emit a loud low-frequency sound source at 939 m deep and measure its travel speed to a receiver located 5,000 km away. The measurement was to be done in 1996 and again in 2006. Since changes in sound speed are caused primarily by changes in the temperature of the ocean, the idea was to measure the rate at which ocean temperatures were increasing to get an estimate of the rate of climate change.

We set out to determine if the loud low-frequency ATOC sound source affected the behavior of elephant seals. We attached acoustic data loggers to juvenile elephant seals and released them at a location in central California where they had to swim over the ATOC sound source to return home to Año Nuevo (Fletcher et al. 1996; Burgess et al. 1998; Costa et al. 1998; Oliver et al. 1998). The sound levels received on the seals was sufficiently low that they were barely audible to the seals; we observed no changes in the diving pattern; we noted only subtle changes in descent and ascent rates. We concluded that the sound source had minimal effects on the diving behavior of the seals.

Whether migrating elephant seals, which cross noisy ship traffic lanes, are affected by the low-frequency sounds made by the tankers is unknown. Gross changes in migration routes or the diving pattern are not evident as they cross these noisy shipping lanes.

How important is the auditory sense to seafaring elephant seals? Is it any help in capturing prey? In navigation? In avoiding predators? In avoiding ship traffic?

13 Comparisons, Unsolved Mysteries, and Conclusions

Sometimes the closer you are to the truth, the harder it is to see.

Jacqueline Simon Gunn

This review shows that elephant seals stand out in many respects. Several salient aspects of their biology serve as a springboard for comparing them with other large mammals and for putting their behavior in perspective.

Long-term Studies

Most of what we know about elephant seals comes from long-term study of wild, marked, known-age individuals living in nature as they might have lived hundreds or thousands of years ago. Long-term study is necessary and invaluable for understanding the factors that influence all facets of their natural history, such as the degree of polygyny, sexual dimorphism, reproductive success, and the variables that affect survival. The individuals that survive and reproduce determine life in future generations. Comparisons of the long-term studies between various mammals are most instructive.

The 1960s were breakthrough years in the study of animal behavior, in large part because this is when many long-term studies began, many of which continue to this day. At the time, long-term studies of wild animals were a great departure from previous approaches to studying animals, such as ethology and comparative psychology. The new long-term studies of wild animals focused on individuals. This was especially important during the 1970s as the paradigm shifted from interpreting selection acting on groups to acting on individuals. Sociobiology embodied the neo-Darwinian approach. This new emphasis focused on the forces that shaped behavior, its function, the operation of natural selection, the factors associated with survival and who won in sexual competition, and how this influenced dispersal and the formation of new colonies and the adaptation of populations to the environment. Knowing the fundamental natural history of a species is important. For example, ecological studies that address the interactions between multiple species require, first of all, deep knowledge of each individual species.

The study of elephant seals at Año Nuevo began in 1962, at about the same time as Jane Goodall's classic studies of chimpanzees (Lawick-Goodall 1967, 1971), and the colony has been monitored continuously for 58 years. Similar lengthy studies have been conducted on southern elephant seals (SES) at various sites in the southern hemisphere (see Chapter 9). These studies are comparable with other long-term studies of large mammals, such as: (1) red deer on the Isle of Rum since 1972 (49 years) (Clutton-Brock et al. 1982); (2) reindeer in Finland since 1969 (51 years) (Weladji et al. 2006); (3) cheetahs of the Serengeti, Tanzania, since 1974 (25–36 years)(Caro 1994); (4) black-tailed prairie dogs in South Dakota from 1975 to 1989 (14 or more years)(Hoogland 1995); and (5) African elephants in Amboseli National Park, southern Kenya, from 1972 (48 years)(Lee et al. 2016). Of course, these long-term studies vary from each other in many respects. The ideal is a situation where the animals are wild and free from human interference. Nevertheless, some populations are initiated on islands by humans, restricting the animal's range and affecting dispersal; in some others, individuals are confined at times, or slaughtered to control numbers, or poached; and in others, tourists may impact the behavior of individuals. In brief, the "wildness" of long-term studies varies. During studies of elephant seals at Año Nuevo, select harems were observed from a distance of 25 m by visitors on guided tours, and in isolated harems on the periphery of the colony, some seals were immobilized for attachment of dive recorders or other procedures; these operations had little impact on the seals, which are robust against disturbance (Le Boeuf 1995; Le Boeuf and Campagna 2013).

Polygyny and Sexual Dimorphism

To put elephant seal social behavior in perspective, it helps to consider the fundamental mating systems observed in nature. Mating systems show how individuals act to maximize their reproductive success (Davies 1991). Three main categories apply to animals and birds: monogamy, polygyny, and polyandry (Emlen and Oring 1977). The systems vary according to the energy required to wean or fledge the young and the parents that provide it.

Monogamy occurs in animals where both the male and the female are needed to nurture the young to independence; this is the most common system in birds where both parents can feed the young. Monogamy is rare in mammals (Kleiman 1977) but is common in modern humans and appears in several terrestrial carnivores where males help feed pups (MacDonald 1983). Monogamy prevails when females are widely dispersed, such as nocturnal primates that live in forests (Rutberg 1983); this gives scant opportunity for males to mate with more than one female. Experience gained by avian couples that mate together annually increases their lifetime productivity compared to birds that pair-bond with a new mate each year (Orians 1969).

Polyandry is rare in nature but is observed in bird species where life is tough and unpredictable, requiring a female to get the help of several males to nurture the young. In some birds, such as the spotted sandpiper (Oring and Knudson 1973; Oring and Lank 1982), females lay a clutch of eggs that one male incubates, then she lays another clutch for another male, and so on. This is her best hedge against an uncertain future, such as a storm or a flood, that can destroy a clutch suddenly and require an expensive and risky restart.

Polygyny is the most commonly observed mating system in mammals (Trivers 1972). This includes most ungulates, rodents, primates, and carnivores (e.g., Hall 1968; Downhower and Armitage 1971; Franklin et al. 1975; Hoogland 1995). These mating systems are determined in large part by two classes of phenomena, phylogenetic inertia in parental investment and environmental variables. When females do all the parental care, this frees males to concentrate on mating, and when there is a paucity of breeding sites free from predators, such as islands for seals, females are forced to clump. This provides a male with the opportunity to mate with multiple females. This explains, in large part, why elephant seals are extremely polygynous.

Polygyny means that the variation in reproductive success is greater in males than in females.

All three of these mating systems have been observed in our own species in historical times, notably among the Hindus in parts of India (Samuelson 1890). The system that prevails depends on economics; monogamy is most prevalent, but polygamy is expected of the wealthy because these males have the resources to support many wives, while the poor have to resort to polyandry because a female requires the help of multiple males.

Measures of male reproductive success are useful for distinguishing mating systems. A male's reproductive success (RS) is the number of pups sired. RS is usually estimated from observing copulations, which yields a measure of mating success. When a female mates with only one male, as in many monogamous situations, or in territories of otariids where the female mates once and this is usually with the territorial male, it is reasonable to assume that the union results in insemination; the male sires the resulting pup and his mating success reflects his reproductive success. In polygynous systems – such as in elephant seals – where females may mate with multiple males, mating success is less directly related to reproductive success. Ideally, DNA fingerprinting would provide a direct measure of parentage to validate observations of mating success (Boness 1991), but for various reasons, this approach has proven difficult to implement. It is also not possible at this time for northern elephant seals (NES) because they lack sufficient genetic variability to determine paternity. The most common method of estimating the degree of polygyny is mating success. This is easy to document on land but not in the water. In Weddell seals, *Leptonychotes weddelli*, that breed under ice, G. L. Kooyman (1981) found that only one copulation was ever observed after having studied these animals for 15 years. Hill (1987) applied colored grease to males and showed that some males came into physical contact with as many as eight females, from which it was suggested that polygyny was moderate in the species. Our present estimates of the degree of polygyny are coarse and could be improved by genetic analysis.

All land-breeding pinnipeds are polygynous (Le Boeuf 1991). This includes all of the sea lions and fur seals (Boness 1991), most phocids such as the grey seal, *Halichoerus grypus*, and the walrus. Species, however, vary from being extremely polygynous to only slightly polygynous (Boness 1991; Le Boeuf 1991). Elephant seals stand out as being extremely polygynous and sexually dimorphic in comparison with other large mammals. Next to elephant seals, the highest level of polygyny is observed in northern fur seals, *Callorhinus ursinus*, also called the Alaska fur seal (Bartholomew and Hoel 1953). Their territories contain on average 40–50 females, but up to a maximum of 100 females, and the territorial male may mate with all of them in a single breeding season (Bartholomew and Hoel 1953; Peterson 1965; Gentry 1998). Copulations are not interrupted by neighboring males because the territorial boundaries have been defined by threats and fighting in advance. Bulls that return in successive years tend to occupy the same territories each year. It follows that these animals are extremely sexually dimorphic, with males being about four times heavier than adult females. South African fur seals, *Arctocephalus pusillus*, may have a similar level of polygyny (David 1987). In other otariids, the maximum number of females mated by a territorial male during a single breeding season is 30 or less, and males are on average three times heavier than females. In all of these species, fertilizable females outnumber sexually active males (Boness 1991). Territories in California sea lion, *Zalophus californianus*, which breed on open sandy beaches, are more fluid and less fixed than other otariids and may include elements of lekking (Heath and Francis 1983, 1987). Southern sea lions, *Otaria byronia*, are territorial in some areas and defend females in other areas (Campagna and Le Boeuf 1988, Campagna et al. 1988), demonstrating that the environmental context influences the mating strategies of males. Data on lifetime reproductive success, however, are not available in these species owing to the difficulty in tracking individuals from one year to the next and throughout life (Gentry 1998).

Moderate polygyny is expected in most otariids because estrus synchrony is brief (90% of females are receptive within a short three- to four-week period, compared to 2½ months for elephant seals). Moreover, males are not capital breeders and must fast to maintain control over females in their territory. The southern sea lion, Galapagos fur seal, and Hooker's sea lion have low levels of polygyny, as do phocids such as the Weddell seal and the grey seal (Boness and James 1979; Hill 1987). As with elephant seals (Le Boeuf and Reiter 1988), variance in lifetime reproductive success of most pinnipeds is expected to be greater in males than females.

Little is known about phocids that breed on ice or in the water. The pagophilic seals are harp, (*Phoca groenlandicus*), ringed (*P. hispida*), ribbon (*P. fasciata*), bearded (*Erignathus barbatus*), hooded (*Cystophora cristata*), Baikal (*P. sibirica*), Caspian (*P. Caspica*), and larga seal (*P. largha*) in the northern hemisphere and Ross (*Ommatophoca rossi*), leopard (*Hydrurga leptonyx*), and crabeater seal (*Lobodon carcinophagus*) in the southern hemisphere, plus the monk seal (*Monachus monachus* and *M. schauinslandi*), which mates in temperate waters, and the harbour seal (*P. vitulina*), which mates in the water near land or ice (Kovacs 1989). These seals are not found in large groups on expansive ice habitats (Stirling 1975); they are widely distributed from each other, which permits only slight polygyny or facultative monogamy, meaning the male remains with a female until she comes into estrus and they mate and then he moves on to another female, that is, another short-lived conjugal relationship.

Female walruses, *Odobenus rosmarus*, in the Bering Sea congregate in groups on ice floes. Males display in the water on the perimeter competing with each other to attract receptive females (Fay et al. 1984). This resembles lekking behavior in ruminants. Lekking is observed when males cannot defend females directly or defend a resource needed by females. Male Weddell seals hold underwater territories, or maritories, near breathing holes used by females (Kooyman 1981). It is not clear whether males intercept females in the water or whether females shop for the "best" male, that is, which sex is in control. A similar arrangement may occur in harp seals near leads in the ice where females haul out.

The general rule, however, is that the mating system of all otariids is resource defense polygyny; males that hold territories mate with multiple females who use these territories to give birth and nurse their pups. Females outnumber males during the breeding season, and males are larger than females.

In the female or harem defense mating system, males defend females directly, and the prospects for polygyny are especially high when females clump together and estrus is moderately synchronized, such as over a period of about two months. The prime exponents of female defense polygyny in pinnipeds are grey seals and elephant seals. The strategies of males across these genera, however, are quite different and so are the consequences.

Approximately 8,000 grey seals breed annually off the east coast of Britain on various islands in Scotland and Wales (Anderson et al. 1975; Hammond et al. 1993). The mating system is unusual in many respects, and there has been debate on whether the mating system is monogamous, territorial, or based on dominance hierarchies, or whether the differences are due to location. Boness and James (1979) describe the mating system on Sable Island in Nova Scotia. Females gather in large groups during the breeding season but keep a distance of about 4.5 m from each other. Males arrive at the same time, follow the females inland, and males compete to gain a position near one or more females. Males defend their position from other males but there is no dominance hierarchy. The tenure of males near females averages 17 days with a range of 2–30 days. Male tenure is highly and positively correlated with mating success. It is estimated that a tenured male achieves six to nine copulations during a breeding season. Although non-tenured males attempt to mate with females when they wean their pups and attempt to return to sea, few are successful. In summary, the key strategy of males is they are "resolute in defense of their female consorts." The larger size of males and their secondary sexual characteristics, which include a long snout (hence, the nickname horseheads) and a rugose neck shield, are consistent with a history of polygyny.

The degree of polygyny in elephant seals is off scale compared to other phocids, otariids, and most terrestrial mammals. In both NES and SES, the alpha and most dominant males mate with 50–100 females in breeding season and perhaps up to 200 or 300 females in life (Chapter 7). In NES, mating yields good estimates of lifetime reproductive success, which is not available for other pinnipeds and available for only a few terrestrial mammals such as red deer (Clutton-Brock et al. 1982, 1988) and prairie dogs (Hoogland 1995).

Numerous studies show that the SES are as polygynous and sexually dimorphic as the NES (Laws 1956; Carrick et al. 1962; Bryden 1972, MCann 1981, 1982; McCann et al. 1989). This is true of many other aspects of their social behavior and physiology (Laws 1953; Le Boeuf and Briggs 1977; Hindell and Burton 1987; McCann 1981, 1985; McCann et al. 1989; Hindell 1991; Campagna et al. 1992; Le Boeuf and Laws 1994; Deutsch et al. 1994; Arnbom 1994; Slip 1997). For example, the annual cycle of breeding and molting on land and the periods spent at sea are similar except that the time scale is shifted; peak breeding is late January in the north and mid-October in the south, the coldest months of the year in both hemispheres. The mating strategies of males are similar in both species; harem size varies with location in both species. Pups are weaned in 23 days in SES and 27 days in NES; NES pups gain 8% more mass per day than SES pups. Maternal energy expenditure is similar for the two species (Arnbom 1994; Crocker et al. 2001) On average, SES females lose 35% of their post-partum mass during lactation, which approximates the 37% lost by NES females. In both species, females are more likely to skip breeding the year following giving birth for the first time, and size at weaning is positively correlated with survival. Pup mortality on the rookery is higher in NES (13%–40%) than SES (2%–16%) and this appears to be due to the higher density of harems in the north. On average, about 50% of the pups of both species die at sea during the first year.

The differences between NES and SES are slight, considering the different environments inhabited by the two species. The context for breeding varies across the numerous rookeries where SES breeds – the substrate varies from sand to pebbles to ice. It is colder in Antarctic waters than in the eastern Pacific where NES breeds and forages. The notable difference in the threat calls of males of both species (Le Boeuf and Petrinovich 1974a) may simply reflect that behavior is most malleable and changeable to the circumstances, and these calls change quickly over time (Casey et al. 2018). In any case, this difference seems inconsequential because the effect of the threat calls to competitors is the same in both species.

Sexual dimorphism, with males being larger than females, is positively correlated with polygyny in most mammals, as observed by Jarman (1974), Clutton-Brock et al. (1977); Alexander et al. (1979); Payne (1984). For example, it is common among mammals that vary greatly in size, from ground squirrels and marmots to African lions and elephants (Morton and Sherman 1978; Armitage 1981; McClean and Towns 1981; Michener 1984; Barash 1989; Packer et al. 1998; Lee et al. 2016). Among adult prairie dogs, males are 10%–15% heavier than females throughout the year (Hoogland 1995). The difference in size between the sexes in elephant seals, however, is approximately twice as great as that of any other pinniped and greater than that of any terrestrial mammal. The mass of a three-year-old female elephant seal at the beginning of the breeding season (who will be mating for the first time) is 360 kg while that of an adult male is 1,814 kg (Deutsch et al. 1994). On average, the male is five times heavier than the female with whom he will mate; in extreme cases the male is ten times heavier than his female mate!

A final note of caution is warranted about these comparisons. The data on SES were obtained from several island and mainland rookeries around the Southern Ocean in a wide belt on both sides of the Antarctic Convergence, principally from South Georgia, Macquarie, Heard, Kerguelen, Gough, Falklands, and Patagonia, Argentina. Most of the data on NES derive from studies on one rookery, the island and mainland at Año Nuevo State Park in central California.

Survival, Lifespan, and Senescence

Most elephant seals die before reaching reproductive age. Their survivorship curve is what Wilson and Bossert (1971) call Type III, the most common in all of nature. The plot of survivors as a function of age plummets early and then levels off with the survivors having an increasingly good chance of reaching maturity and with some living a long life. For comparison with other mammals, we want to know who survives and why and the pattern of reproduction in the survivors throughout life. How important is lifespan in determining lifetime reproductive success?

Females outlive males in most mammals. For 229 species, Xirocostas et al. (2010) found that the homogametic sex, on average, live 17.6% longer than the heterogametic sex; it is the opposite in birds. For 101 species, Lemaitre et al. (2020) found that female lifespan, on average, is 18.6% longer than conspecific males. In humans, the female advantage is 7.8%. In NES the maximum lifespan of females is 64.2% greater than that of males (23 vs. 14 years, respectively). Lifespan for known-age males is based on never having seen a male live past 14 years in five decades of monitoring the Año Nuevo rookery. This is also the maximum lifespan of male SES. The maximum lifespan for known-age female NES is 23 years. This is based on 1,934 of 7,735 female pups weaned that survived to breed at least once. The discrepancy in lifespan between the sexes in elephant seals is one of the largest in mammals.

Among elephant seals, longevity is positively correlated with the lifetime reproductive success of females. Since females breed annually producing a single pup, females that live long have the most opportunities to reproduce. Long-lived females are the most fit, capable of producing up to 20 pups in a long life. How common is that? Longevity has a major influence on lifetime reproductive success for females of many species such as lesser snow geese, splendid fair-wrens, northern fulmars, African lions, African elephants, vervet monkeys, and prairie dogs (Cheney et al. 1988; Cooke and Rockwell 1988; Ollason and Dunnet 1988; Packer et al. 1988; Rowley and Russell 1990; Hoogland 1995; Lee et al. 2016). Although males of all polygynous species have higher variance in reproductive success than females, the higher lifetime reproductive success of long-lived females – the supermoms in elephant seals – help to narrow this gap.

Elephant seals are similar in many respects to that other prominent, long-nosed, mammoth, the African elephant. In one study, less than 30% of the population of Amboseli elephants survived past the onset of fertility decline, which begins at age 40, but those that lived past this age reaped higher rates of reproduction throughout their lives (Lee et al. 2016). Longevity is associated with high reproductive rates in elephant seals, elephants, humans (Thomas et al. 2000; Hawkes and Smith 2010), wild chimpanzees (Thompson et al. 2007), and a range of birds and mammals (Rebke et al. 2010). Long-lived females are fit. Moreover, early breeding elephants, like in a variety of species (Nussey et al. 2006; van de Pol and Verhulst 2006; Robinson et al. 2012; Desprez et al. 2014), show higher reproductive rates compared to late breeders. In elephant seals, females with the highest lifetime reproductive success start breeding early, and they do it successfully, but many of the other females that try this fail to wean their first pups.

Both female Amboseli elephants and elephant seals exhibit the classical mammalian pattern of an age-specific decline in reproductive rate. When elephants start showing signs of senescence, they grandmother the offspring of family members for about 15 years in addition to continuing to reproduce. Most old female elephant seals continue to breed until they die; they do not act as grandmothers, and reproductive senescence is moderate and only becomes apparent in extreme old age, the year or two before they die.

Males? Longevity is less important for males than females. A dominant alpha male can reap great reproductive success in a single breeding season, or even more, if his tenure as alpha continues for multiple years. Males surpass the lifetime reproductive success of all females in one good breeding season. This assumes that mating success is positively correlated with paternity. We documented that one male, ADR – who is certain to be elected to the elephant seal Hall of Fame – dominated mating in the same harem at Año Nuevo for four consecutive years (Le Boeuf 1974). This is rare. The important aspect of longevity for a male is to live to peak breeding years, ages 9–12. Few manage this. In one study where we tracked known-age males until they died, only 4.4% of male pups weaned reached age 9; only 2.2% lived to age 12 (Le Boeuf and Reiter 1988). Living longer than age 12 doesn't help males as it does with females. A decrease in competitiveness is evident at age 13, a sign of senescence. Males are clearly "over the hill" at age 14 and none live beyond this age.

Aquatic Behavior

The diving, foraging, and migratory pattern at sea is exceptional for both species of elephant seals. At-sea behavior of both species has been studied intensively using a variety of diving instruments attached to the seals that monitor multiple variables such as location in real time, dive depth and duration, dive types that reflect the function of dives, swim speed, body position, water temperature and salinity, prey caught and consumed, and more. For the seals in the southern hemisphere, emphasis is placed on location of feeding, their prey, and estimating Southern Ocean primary production. The emphasis on the northern species has run the gamut from detailed aspects of diving and its function and physiological basis to development and sex differences in diving behavior and location of diving and the physiology that enables long and deep diving in pursuit of prey.

The elephant seals dive deeper and longer than other seals, sea lions, and the walrus. They differ from other diving seals in that the time spent at the surface between dives is less than three minutes, even following long, deep dives. This suggests that elephant seals never rely on anaerobic metabolism, which would require time to recover at the surface and get rid of anaerobic metabolites before being able to dive again. The sexes differ in the types of dives displayed, with males showing a preponderance of benthic dives, most of which are probably depth-limited by the ocean bottom, and females favoring pelagic dives in the open ocean (Hindell et al. 1991; Slip 1997; Le Boeuf et al. 1992; Le Boeuf and Naito, in press). Both sexes may exhibit the less preferred dives depending on where they are foraging. The sex differences in dive types displayed reflect the behavior and distribution of the prey each sex pursues – benthic prey for males and pelagic prey for females (see Chapter 9).

Adults of both species seem to have no difficulty in finding sufficient prey to build up body stores, stored in blubber, which help maintenance and reproduction while they fast on land. The exception is that some adult female NES have difficulty foraging in severe El Niño years (Le Boeuf and Crocker 2005; Crocker et al. 2006). Another exception is that pups of both species going to sea for the first time suffer high mortality, which may be due to difficulty in finding food.

The dives of elephant seals are deeper and longer than those of most whales and dolphins, the exceptions being beaked whales and sperm whales. The comparisons are more meaningful if we provide context. First, the order Cetacea is divided into two sub-orders – odontocetes, or toothed whales, and mysticetes, the baleen whales (Gaskin 1982). Odontocetes are divided into three super-families, the sperm whales, the beaked whales, and the delphinids, which includes the dolphins, killer whales, and pilot whales. Mysticetes include two families, the baleen whales and the rorquals. Second, the methods used to document diving in seals differ from that used to study whales and dolphins. The most frequently used dive recorders applied to seals are archival; at minimum, they record the depth of all dives and surface intervals in great detail for the long periods that the seals are at sea, which is up to eight months in gestating females (Hindell et al. 1991; Le Boeuf et al. 1993; 2000a). The recorders, however, must be recovered to get the data. In most cases, this is done when the seals return, reliably, to the rookery.

Most recorders applied to whales and dolphins (Hooker and Baird 2001) remain on the animals for only a few hours (Hooker and Baird 1999; Croll et al. 2001; Baumgartner and Mate 2003; Baird et al. 2006; Simon et al. 2009) and must be recovered at sea or the data collected must be up-linked to a satellite. Satellite data are not as detailed and precise as archival data. Recent deployments of satellite-linked tags remain on the whales for months (e.g., Shearer et al. 2019). Some studies of beaked whales, such as those of Quick et al. (2020) and Shearer et al. (2019), restrict analysis to deep, long dives and treat shallow dives separately. The different modes of analysis, such as using binned data from satellite tags, make comparisons between the seals and whales difficult.

Despite the different methods used and the duration of recordings, two statistics permit reasonable comparisons between the seals and cetaceans, the deepest dive and the longest dive duration recorded. The deepest dive recorded for a cetacean is

2,992 m for Cuvier's beaked whale, *Ziphius cavirostris* (Schorr et al. 2014). Baird's beaked whale, *Berardius bairdii*, recorded a dive of 1,777 m (Minamikawa et al. 2007). Maximum dive depths of sperm whales have increased over the years: 900 m, 1,138 m (Heezen 1957), 1,140 m (Lockyer 1977), and 1,185 m or possibly 2,035 m (Watkins et al. 1993). A narwhal, *Monodon monoceros*, may dive to 1,500 m or more but most dives are between 200 and 400 m (Laidre et al. 2003). The maximum dive depths of all other whales and dolphins are less than 1,700 m. The maximum depths of bowhead whales is 538 m (Heide-Jørgensen et al. 2013), deeper than any other mysticete whale, but most dives for bowheads are much shallower. The foraging dives of North Atlantic Right whales are between 80 and 175 m (Baumgartner and Mate 2003). Blue whales dive to a maximum depth of 173 m and fin whales to 180 m, but on average, blue whales dive to 140 m and fin whales to 97.9 m when foraging (Croll et al. 2001).

In comparison, the maximum dive depth recorded for NES is 1,735 m (Robinson et al. 2012) and 1,872 m (D.P. Costa, pers. comm.) and 2,165 m for SES (McIntyre et al. 2010). When it comes to deep diving, the elephant seals are in elite company.

The longest dive recorded for Cuvier's beaked whales lasted 137.5 minutes (Schorr et al. 2014). An exceptionally long dive of 222 minutes may have been a response to a Navy mid-frequency active sonar signal (Quick et al. 2020). A harpooned northern bottlenose whale, *Hyperoodon ampullatus*, remained submerged for 120 minutes (Gray 1882). The longest dive recorded for a Baird's beaked whale is 64.4 minutes (Minamikawa et al. 2007). The longest dives for baleen and rorqual whales are shorter in duration: about 40 minutes for bowheads, 14.7 minutes for blue whales, and 16.9 minutes for fin whales (Croll et al. 2001; Heide-Jorgensen et al. 2013).

In comparison, a dive lasting 120 minutes was recorded for a female SES (Hindell et al. 1991) and 103 minutes for a female NES (Hassrick et al. 2010), and 116 minutes for another female NES (D.P. Costa, pers. comm.) Here, too, elephant seals are in company with the longest diving cetacean, the Ziphiidae.

Hooker and Baird (1999) point out that the maximum depths and dive durations recorded may not be the best representations of the diving behavior of an animal (Hooker and Baird 1999). This is a good point but routine dive depths and durations are misleading as well when some are taken from a few hours of diving or when deep, long dives are treated separately, as with many whales, while others such as those from elephant seals represent all dives during several months at sea.

With this stipulation in mind, the routine diving behavior of elephant seals reveals just how aquatic they are. During the months at sea, their depth of dives ranges from 400 to 700 m (Le Boeuf et al. 1988; DeLong and Stewart 1991; Hindell et al. 1991; McIntyre et al. 2010). The overall mean dive depth of adult female NES is 516 m, but since the depth of dives varies with available light, this breaks down into 619 m by day and 456 m by night. Most dives last about 20 minutes with only about 2 minutes at the surface between dives; this brief surface interval holds even after the deepest and longest dives. Consequently, elephant seals of both species spend 90%–91% of their total time at sea underwater. It would be revealing to compare seals and whales on time under-water, but this statistic is either difficult to obtain or is rarely reported in studies of cetacean diving behavior. Elephant seals are exceptional divers that have few peers.

Unsolved Mysteries

Mystery 1: Death at sea. Although we know much about the natural history of elephant seals, some things remain a mystery. The biggest puzzle for the next generation of sea-elephantologists is death at sea. Most deaths occur at sea. This is obvious from the low incidence and absence of deaths on the rookery and on land. The conclusion from five decades of observing elephant seals during the breeding season, the molt, and off-season haul-outs is that the mortality of juveniles and adults of both sexes on land is rare. Therefore, the vast majority of deaths occurs at sea. Of course there are exceptions; suckling pups and weanlings die in the harems or in the dunes nearby; the causes of these deaths were discussed in Chapter 6. In addition, a few adult females are killed inadvertently by males attempting to mate with them as they leave the harem to go to sea to forage (Chapter 8). Very rarely, a male is killed in a fight. I saw just one or two in 50 years of observations.

Consequently, we must examine the two annual foraging trips at sea to explain the high mortality before the seals reach breeding age, as well as the circumstances that prevent most breeders from surviving to old age. Likely causes of death at sea are predation, failure to find sufficient food, entrapment in nets or debris, and ship strikes.

We know that white sharks, *Carcharodon carcharias*, kill elephant seals of all ages and both sexes near the Año Nuevo and Farallon Island rookeries (Ainley et al. 1981; Tricas and McCosker 1984; Le Boeuf and Crocker 1996; Klimley et al. 2001; Boustany et al. 2002; Le Boeuf 2004) (Figure 13.1). Approximately a dozen or more white sharks, which are known and identifiable, hunt and feed on the seals in the waters near Año Nuevo Island from about October through about mid-December and return to forage here year after year (Figure 13.2). The number of seals taken and the significance of the predation, however, is unknown. Cookie cutter sharks, *Isistius* (Le Boeuf et al. 1987), do not kill seals but parasitize them by biting out chunks of skin and blubber the size of tennis balls. Even less is known about the predation of killer whales, *Orca orcinus,* on elephant seals except that it occurs and has been observed. Similarly, some seals get struck by ships or ship propellers (Figure 13.1.d). Some are entrapped and die in fishing nets or discarded fishing gear during their migrations in the northeastern Pacific Ocean (Le Boeuf et al. 2000a; Jepsen and Nico de Bruyn 2019), but it is not clear how many die this way.

We know that some adult females have difficulty finding food during severe El Niño events (Huber et al. 1991; Le Boeuf and Reiter 1991; Stewart and Yochem 1991; Le Boeuf and Crocker 2005; Crocker et al. 2006; Adachi et al. 2014). They spend more time foraging, time in prey patches is reduced, travel time between prey patches increases, mass gain rate decreases substantially, and some females gain no weight at

Figure 13.1 Injuries observed. Upper left: an adult female that was recently bitten by a white shark. Upper right: two large chunks of flesh taken out of the side of a dead adult male elephant seal by a white shark. Lower left: a dead subadult male showing signs of having been attacked and killed by one or more sharks. Lower right: an adult female with pup with a large wound on her neck and over her shoulder that appears to have been caused by a ship's propeller. (A black and white version of this figure will appear in some formats. For the color version, please refer to the plate section.)

Figure 13.2 A large, dead white shark photographed shortly after it was washed ashore at Año Nuevo State Park. Photograph by Kenneth Parker.

all. Poor foraging may be because the prey are disrupted by the warm waters, their normal distribution changes, and the seals fail to locate them. The upshot is that poor foraging of mothers reduces the weaning mass of pups and lowers the probability of their survival. It is notable that large, old mothers are less affected than other females in poor foraging years. It is not clear if mortality is affected directly by the inability of seals to gain sufficient nourishment.

The long lives of some females stand out in contrast with the high mortality rate of young females that don't reach breeding age and the young female survivors that live to breeding age but breed only a few times before they die (Le Boeuf et al. 2019). Evidently, the long-lived, multiparous females – the supermoms – are better at avoiding the pitfalls at sea that befalls most other females. Why? We don't know but there are promising leads to follow up on that might shed light on this matter.

Researchers at the University of California at Santa Cruz have tracked over 600 adult females using a variety of dive recorders, satellite tags, GPS tags, and acoustic tags, which measure diving behavior, location, speed of travel, and so forth in the last few decades (e.g., Le Boeuf et al. 2000a; Costa et al. 2012; Robinson et al. 2012). This database affords the possibility of comparing the tracks and behavior of supermoms to other females. Do they forage in the same place, a safer place? If an instrument stops abruptly, is it due to instrument or battery failure or is it because the seal died. Where do the "stops" occur? These are just a few questions that can be addressed to help elucidate why, when, and where the seals die at sea. New additions are being added to the arsenal every year. Some developments promise being able to distinguish the moment of death and the behavior and location just before death. Addressing the causes of death at sea is a new frontier in research on elephant seals.

In a recent study by Kienle (2019), 32 adult males and 152 adult females were tracked with satellite transmitters and time–depth recorders. Males had a significantly lower survival rate at sea (56.4%) than females (87.3%), and this pattern prevailed across all years and both annual foraging trips. The diving instruments on 16 males and 22 females stopped transmitting *en route*. It is not clear what caused the stops but some of them may have been because the seal died. The instruments of 12 males stopped transmitting near the coast where they forage and 4 at sites at sea along the migratory paths. The instruments of 2 out of the 12 males stopped transmitting near the Aleutian Islands. In contrast, the instruments of 3 out of the 22 females stopped near the coast and the rest stopped along their migration routes.

Mystery 2: Virgin females. The other puzzle that has long intrigued some of us is when, where, and with whom do virgin females mate. We know that most females give birth for the first time at age 4; a substantial number are primiparous at age 3; and rarely a female gives birth at age 2. With multiparous females there is a year between conception and giving birth. If primiparous females operate on the same schedule as multiparous females, they mate the year before they give birth for the first time; that is, a female mates at age 3 to give birth at age 4, mates at age 2 to give birth at age 3, and mates at 1 year of age to give birth at age 2. The problem is we rarely see non-pregnant one-, two-, or three-year-old females mating in harems during the breeding season. We might have seen a handful mating at the beginning of the breeding season over a decade, but these are dubious. Judging age from size is unreliable, and if the putative virgins are tagged, their tags must be read to confirm their age. Virgins might be mating in the water, but we do not observe this. If this is the case, who are the males inseminating them? Most of the males in competition for females are on the breeding beaches near harems and they rarely enter the water except for brief fights. Are young males, which cannot compete for females in or near harems, patrolling the waters offshore for virgin females?

Multiparous females get inseminated during the winter breeding season but implantation of the fertilized egg to the uterine wall is delayed until the end of the molt in late Spring, when the eight-month gestation period begins. Is it possible that virgin females mate during the molt and gestation starts immediately, that is, they do not delay implantation during the first pregnancy? We don't see virgins mating on land or in the water during the molt.

One thing seems to be clear. Virgin females are not mating with dominant males around which the whole social/sexual structure is organized. This means they must be mating with young pubertal males or young subadults, that is, males of undemonstrated fitness. Genetic studies could provide answers if the lack of genetic variability could be solved or bypassed or the question phrased in such a way that differentiating specific individuals would not be necessary. For example, it might be a question of what proportion of the pups produced by primiparous females were sired by the dominant males (or all males) in residence the previous year? This assumes, of course, that we can obtain minute tissue samples from all the males and the mothers and their pups, which is feasible.

Imagine that we find that the virgins were inseminated by young pubertal males. So what? About 50% of females giving birth for the first time fail to wean their pups. Those pups that survive are smaller than average because their mothers were smaller than older females, and therefore, their pups have a lower probability of surviving and breeding. My point is that virgin matings may not be so important. Like in many animals, first times are practice for later. And practice improves performance.

More questions. The at-sea behavior of elephant seals is an exceptionally fertile venue for additional behavioral and physiological studies. How do the seals deal with high pressure at depth and the pressure changes while diving? What allows these seals to dive for an hour or more and yet require only a minute or two before diving again? How much time at sea is spent sleeping, and when does sleep occur during the course of a dive? Do males feed during long migrations to foraging destinations? What are the clues to navigation to the same place in successive years? How are the seals, and especially the males, capturing prey during benthic dives? What is the principal prey of males during benthic dives? How will these seals deal with warming waters and climate change?

Many questions remain to be solved about the physiology of diving. According to Blix (2018) we need to know more about (1) how cells shut down to save energy when oxygen is in short supply and how neuron metabolism is organized; (2) why surface intervals do not vary with the duration of preceding or succeeding dives in elephant seals; and (3) how lactate is eliminated and how the toxic effects of high partial pressures of nitrogen and oxygen at depth are avoided.

Do males that mate with departing females sire pups? Or are these females already inseminated by alpha or other dominant males in the harem? Do bulls in the harem that mate frequently lose the ability to inseminate females? Are the progeny of alpha bulls and supermoms more likely to become alphas and supermoms themselves? Questions lead to more questions.

Conclusions

Sports enthusiasts are impressed by records that signify the best performances ever. The record holder for most homeruns in baseball. The football player with the most yards gained or the most touchdowns or the most passes caught. The golfer who won a tournament by beating his competitors by 14 strokes. Athletes like Pele, Ronaldo, and Messi are idolized for the most goals made in a lifetime of playing soccer, and Curry and Thompson are lauded for the most three-point goals in basketball. And Alex Morgan because she is the best female soccer player today. Well, I think so.

There is similar regard for superlative performances by animals. I guess it is the way we are built. Cheetahs fascinate us because they are the fastest terrestrial carnivore. We see pythons as physiological wonders because they can swallow prey as large as deer, cattle, or a human. Bats are a marvel to us because they can "see" in the dark using echolocation. And so on.

Elephant seals excel not in one single impressive feat but in multiple respects: the species is exceptionally polygynous and sexually dimorphic; females fast from food and water throughout the one-month lactation period, and males fast for over three months while actively fighting and competing to mate; the energy exchange between the mother and pup is 50%, and milk exceeds 50% fat, allowing pups to triple their mass in four weeks; these seals hold their breath when sleeping on land; they dive deeper and longer than other pinnipeds and most whales and they spend only a minute or two on the surface breathing between dives; 90% of the time at sea is spent underwater; they can stay underwater for up to two hours; they possess some of the highest known mammalian blood volumes and hemoglobin concentrations in nature; they can withstand high pressure in deep waters and rapid changes in high pressure as they move from the surface to a depth of 1.5 km; and after flirting with near extinction, their population has recovered in number and breeding range.

In summary, elephant seals are exceptional, superlative, extremophiles. They push the behavioral and physiological limits of adaptation on land and at sea. We have learned much from them about the evolution of social behavior, the operation of natural selection, and adaptations for making a living at sea. Continuous study of these animals over five decades was possible, in large part, because we could identify and track individuals throughout their lives, the colonies were close by, and accessible, and we could work close to the seals because they were not afraid of us. Knowledge of their at-sea behavior was advanced considerably by the development of diving instruments they carried for months while free-ranging and foraging. We are in awe of what they can do.

References

Adachi, T., Costa, D. P., Robinson, P. W., Peterson, S. H., Yamamichi, M., Naito, Y., and Takahasi, A. (2017). Searching for prey in a three-dimensional environment: hierarchical movements enhance foraging success in northern elephant seals. *Functional Ecology* 31(2), 361–369.

Adachi, T., Maresh, J. L., Robinson, P. W., Peterson, S. H., Naito, Y., Watanabe, Y., and Takahashi, A. (2014). The foraging benefits of being fat in a highly migratory marine mammal. *Proceedings of the Royal Society B* 281, 2014210.

Aguilar de Soto, N., Visser, F., Tyack, P., Alcazar, J., Ruxton, G., Arranz, P., Madsen, P. T., and Johnson, M. (2020). Fear of killer whales drives extreme synchrony in deep diving beaked whales. *Scientific Reports* 10, 13.

Ainley, D. G., Strong, C. S., Huber, H. P., Lewis, T. J., and Morrell, S. (1981). Predation by sharks on pinnipeds at the Farallon Islands. *Fishery Bulletin – U.S. National Oceanic and Atmospheric Administration* 78, 941–945.

Albro, R. R. (1980). Northern elephant seal vs. dogfish. Cetus, J. Whale Museum 21–22.

Alexander, R. D. (1974). The evolution of social behavior. *Annual Review of Ecology and Systematics* 5, 325–383.

Alexander, R. D. (1979). *Darwinism and Human Affairs*. Seattle: University of Washington Press.

Allen, S. G., Peaslee, S. C., and Huber, H. R. (1989). Colonization by northern elephant seals of the Point Reyes Peninsula, California. *Marine Mammal Science* 5, 298–302.

Anderson, S. S., Burton, R. W., and Summers, C. F. (1975). Behaviour of grey seals (*Halichoerus grypus*) during a breeding season at North Rona. *Journal of Zoology, London* 177, 179–195.

Andrews, R. D., Costa, D. P., Le Boeuf, B. J., and Jones, D. R. (2000). Breathing frequencies of northern elephant seals at sea and on land revealed by heart rate spectral analysis. *Respiration Physiology* 123, 71–85.

Andrews, R. D., Jones, D. R., Williams, J. D., Thorson, P. H., Oliver, G. W., Costa, D. P., and Le Boeuf, B. J. (1997). Heart rates of northern elephant seals diving at sea and resting on the beach. *Journal of Experimental Biology* 200, 2083–2095.

Anson, G. (1748). *A Voyage Round the World in Three Years (1740–1744)*. Compiled by R. Walter. London: John & Paul Knapton.

Antonellis, G. A. and Fiscus, C. H. (1980). The pinnipeds of the California current. *California Coop Oceanic Fisheries Investigations Report* 28, 68–78.

Antonellis, G. A., Jr., Fiscus, C. H., Stewart, B. S., and Delong, R. L. (1994). Diet of the northern elephant seal. In B. J. Le Boeuf and R. M. Laws, eds., *Elephant Seals: Population Ecology, Behavior, and Physiology*. Berkeley: University of California Press, pp. 211–223.

Antonellis, G. A., Jr., Lowry, M. S., DeMaster, D. P., and Fiscus, C. H. (1987). Assessing northern elephant seal feeding habits by stomach lavage. *Marine Mammal Science* 3, 308–322.

Aoki, K., Watanabe, Y. Y., Crocker, D. E., Robinson, P. W., Biuw, M., Costa, D. P., Miyazaki, N., Fedak, M. A., and Miller, P. J. (2011). Northern elephant seals adjust gliding and stroking patterns with changes in buoyancy: validation of at-sea metrics of body density. *Journal of Experimental Biology* 214, 2973–2987.

Armitage, K. B. (1981). Sociality as a life-history tactic of ground squirrels. *Oecologia* 48, 36–49.

Arnbom, T. (1994). Maternal investment in male and female offspring in the southern elephant seal. Ph.D. dissertation, Stockholm University, Stockholm.

Asaga, T., Naito, Y., Le Boeuf, B. J., and Sakurai, H. (1994). Functional analysis of dive types of female northern elephant seals. In B. J. Le Boeuf and R. M. Laws, eds., *Elephant Seals: Population Ecology, Behavior, and Physiology*. Berkeley: University of California Press, pp. 310–327.

Aurioles, D., Koch, P. L., and Le Boeuf, B. J. (2006). Differences in foraging location of Mexican and California elephant seals: evidence from stable isotopes in pups. *Marine Mammal Science* 22, 326–338.

Aurioles, D., Le Boeuf, B. J., and Findley, L. (1993). Registros de pinnipedos poco communes para el golfo de California. *Revista Investigación Científica* 1(UABCS), 13–19.

Bailleul, F., Charrassin, J.-B., Monestiez, P., Roquet, F., Biuw, M., and Guinet, C. (2007). Successful foraging zones of southern elephant seals from the Kerguelen Islands in relation to oceanographic conditions. *Philosophical Transactions of the Royal Society B* 362, 2169–2181.

Baird, R. W., Webster, D. L., McSweeney, D. J., Ligon, A. D., Schorr, G. S., and Barlow, J. (2006). Diving behaviour of Cuvier's (*Ziphius cavirostris*) and Blainville's (*Mesoplodon densirostris*) beaked whales in Hawai'i. *Canadian Journal of Zoology* 84, 1120–1128.

Banks, R. C. (1976). Extation. *Science* 191, 1215–1292.

Barash, D. P. (1974). The social behavior of the hoary marmot (*Marmota caligata*). *Animal Behavior* 22, 256–261.

Barash, D. P. (1989). *Marmots: Social Behavior and Ecology*. Stanford, CA: Stanford University Press.

Bartholomew, G. A. (1952). Reproductive and social behavior of the northern elephant seal. *University of California Publications in Zoology* 47, 369–472.

Bartholomew, G. A. (1954). Body temperature and respiratory and heart rates in the northern elephant seal. *Journal of Mammalogy* 35, 211–218.

Bartholomew, G. A. and Collias, N. E. (1962). The role of vocalization in the social behaviour of the northern elephant seal. *Animal Behaviour* 10, 7–14.

Bartholomew, G. A. and Hoel, P. G. (1953). Reproductive behavior of the Alaska fur seal, *Callorhinus ursinus*. *Journal of Mammalogy* 34, 417–436.

Bartholomew, G. A. and Hubbs, C. L. (1960). Populational growth and seasonal movements of the northern elephant seal, *Mirounga angustirostris*. *Mammalia* 24, 313–324.

Baumgartner, M. F. and Mate, B. R. (2003). Summertime foraging ecology of North Atlantic right whales. *Marine Ecology Progress Series* 264, 123–135.

Beirne, P. (2018). Theriocide and homicide. In *Murdering Animals*. Palgrave Studies in Green Criminology. London: Palgrave Macmillan.

Bell, G. (1980). The costs of reproduction and their consequences. *American Naturalist* 116, 45–76.

Bercovitch, F. B. and Berry, P. S. M. (2017). Life expectancy, maximum longevity and lifetime reproductive success in female Thornicroft's giraffe in Zambia. *African Journal of Ecology* 55, 443–450.

Biuw, M., Nøst, O. A., Stien, A., Zhou, Q., Lydersen, C., and Kovacs, K. M. (2010). Effects of hydrographic variability on the spatial, seasonal and diel diving patterns of southern elephant seals in the eastern Weddell Sea. *PLoS ONE* 5, e13816.

Blackwell, S. B. and Le Boeuf, B. J. (1993). Developmental aspects of sleep apnoea in northern elephant seals, *Mirounga angustirostris. Journal of Zoology* 231, 437–447.

Blix, A. S. (2018). Adaptations to deep and prolonged diving in phocid seals. *Journal of Experimental Biology* 221, 1–13.

Block, B. A., Dewar, H., Blackwell, S. B., Williams, T. D., and others. (2001). Migratory movements, depth preferences, and thermal biology of Atlantic bluefin tuna. *Science* 293, 1310–1314.

Bodson, A., Miersch, L., Maiuck, B., and Dehnhardt, G. (2006). Underwater auditory localization by a swimming harbor seal (*Phoca vitulina*). *Journal of the Acoustical Society of America* 120, 1550–1557.

Boehlert, G. W., Costa, D. P., Crocker, D. E., Green, P., O'Brien, T., Levitus, S., and Le Boeuf, B. J. (2001). Autonomous pinniped environmental samplers: using instrumented animals as oceanographic data collectors. *Journal of Atmospheric and Oceanic Technology* 18, 1882–1893.

Boessenecker, R. W. and Churchill, M. (2016). The origin of elephant seals: implications of a fragmentary late Pliocene seal (Phocidae: *Miroungini*) from New Zealand. *New Zealand Journal of Geology and Geophysics* 59, 544–550.

Boness, D. J. (1991). Determinants of mating systems in the *Otariidae* (Pinnipedia). In D. Renouf, ed., *The Behaviour of Pinnipeds*. London: Chapman and Hall, pp. 1–44.

Boness, D. J. and James, H. (1979). Reproductive behaviour of the grey seal (*Halichoerus grypus*) on Sable Island, Nova Scotia. *Journal of Zoology, London* 188, 477–500.

Bonnell, M. L. and Selander, R. K. (1974). Elephant seals: genetic variation and near extinction. *Science* 184, 908–909.

Boustany, A. M., Davis, S. F., Pyle, P., Anderson, S. D., Le Boeuf, B. J., and Block, B. A. (2002). Satellite tagging: expanded niche for white sharks, *Nature* 415, 35–36.

Bradshaw, C., Hindell, M., Best, N., Phillips, K., Wilson, G., and Nichols, P. (2003). You are what you eat: describing the foraging ecology of southern elephant seals (*Mirounga leonina*) using blubber fatty acids. *Proceedings of the Royal Society (B Biological Science)* 270, 1283–1292.

Briggs, K. T. and Morejohn, G. V. (1975). Dentition, cranial morphology, and evolution in elephant seals. *Mammalia* 40, 199–222.

Bryden, M. M. (1972). Body size and composition of elephant seals (*Mirounga leonina*): absolute measurements and estimates from bone dimensions. *Journal of Zoology* 167, 265–267.

Broussard, D. R., Dobson, F. S., and Murie, J. O. (2005). The effects of capital on an income breeder: evidence from female Columbian ground squirrels. *Canadian Journal of Zoology* 83, 546–552.

Burgess, W. C., Tyack, P. L., Le Boeuf, B. J., and Costa, D. P. (1996). Acoustic measurement of cardiac function on northern elephant seals. *The Journal of the Acoustical Society of America* 100, 2709.

Burgess, W. C., Tyack, P. L., Le Boeuf, B. J., and Costa, D. P. (1998). A programmable acoustic recording tag and first results from free-ranging northern elephant seals. *Deep Sea Research Part II: Topical Studies in Oceanography* 45, 1327–1351.

Busch, B. C. (1985). *The War Against the Seals.* Kingston and Montreal, Canada: McGill-Queen's University Press.

Campagna, C. and Le Boeuf, B. J. (1988). Reproductive behaviour of southern sea lions and its effect on mating strategies. *Behaviour* 104, 233–262.

Campagna, C., Dignani, J., Blackwell, S. B., and Marin, M. R. (2001). Detecting bioluminescence with an irradiance time-depth recorder deployed on southern elephant seals. *Marine Mammal Science* 7(2), 402–414.

Campagna, C., Falabella, V. and Lewis, M. (2007). Entanglement of southern elephant seals in squid fishing gear. *Marine Mammal Science*, 23, 414–418.

Campagna, C., Fedak, M. A., and McConnell, B. J. (1999). Post-breeding distribution and diving behavior of adult male southern elephant seals from Patagonia. *Journal of Mammalogy* 80, 1341–1352.

Campagna, C., Le Boeuf, B. J., and Cappozo, H. L. (1988). Group raids: a mating strategy of male southern sea lions. *Behaviour* 105, 224–250.

Campagna, C., Le Boeuf, B. J., Lewis, M., and Bisioli, C. (1992). Equal investment in male and female offspring in southern elephant seals. *Journal of Zoology* 226, 551–561.

Campagna, C., Piola, A. R., Marin, M. R., Lewis, M., Zajaczkovski, U., and Fernández, T. (2007). Deep divers in shallow seas: southern elephant seals on the Patagonian shelf. *Deep-Sea Research I: Oceanographic Research Papers* 54, 1792–1814.

Campagna, C., Quintana, F., Le Boeuf, B., Blackwell, S., and Crocker, D. (1998). Diving behaviour and foraging ecology of female southern elephant seals from Patagonia. *Aquatic Mammals* 24, 1–12.

Carey, F. G. and Scharold, J. V. (1990). Movements of blue sharks (*Prionace glauca*) in depth and course. *Marine Biology* 106, 329–342.

Carlson, K. D. and Le Boeuf, B. J. (1998). Visual pigment sensitivity of the northern elephant seal. Proceedings of the World Marine Mammal Science Conference. Abstract Volume, Monaco, 24 pp.

Caro, T. M. (1994). *Cheetahs of the Serengeti Plains: Group Living in a Social Species.* Chicago, IL: University of Chicago Press.

Croutier, A. L. (1989). *Harem: The World behind the Veil.* New York: Abbeville Press.

Carrick, R. S., Csordas, S. E., and Ingham, S. E. (1962). Studies on the southern elephant seal, Mirounga leonina (L), part 4: breeding and development. Commonwealth Scientific and Industrial Research Organization (C.S.I.R.O.). *Wildlife Research* 7, 161–197.

Casey, C., Reichmuth, C., Costa, D. P., and Le Boeuf, B. (2018). The rise and fall of dialects in northern elephant seals. *Proceedings of the Royal Society B* 285(1892), 20182176.

Cassinello, J. and Alados, C. L. (1996). Female reproductive success in captive *Ammotragus lervia* (Bovidae, Artiodactyula). Study of its components and effects of hierarchy and inbreeding. *Journal of Zoology* 239, 141–153.

Castellini, M. A., Milsom, W. K., Berger, R. J., Costa, D. P., Jones, D. R.. Castellini, J. M., Rea, L. D., Bharma, S., and Harris, M. (1994). Patterns of respiration and heart rate during wakefulness and sleep in elephant seal pups. *American Journal of Physiology* 266, R863–R809.

Cheney, D. L. Seyfarth, R. M., Andelman, S. J., and Lee, P. C. (1988). Reproductive success in vervet monkeys. In T. H. Clutton-Brock, ed., *Reproductive Success: Studies of Individual Variation in Contrasting Breeding Systems.* Chicago, IL: University of Chicago Press, pp. 384–402.

Christenson, T. E. and Le Boeuf, B. J. (1978). Aggression in the female northern elephant seal, *Mirounga angustirostris, Behaviour* 64, 158–172.

Clinton, W. L. and Le Boeuf, B. J. (1993). Sexual selection's effects on male life history and the pattern of male mortality. *Ecology* 74, 1884–1892.

Clutton-Brock, T. H. and Sheldon, B. C. (2010). Individuals and populations: the role of long-term, individual-based studies of animals in ecology and evolutionary biology. *Trends in Ecology and Evolution* 25, 562–573.

Clutton-Brock, T. H., Albon, S. D., and Guinness, F. E. (1984). Maternal dominance, breeding success and birth sex ratios in red deer. *Nature* 308, 358–360.

Clutton-Brock, T. H., Albon, S. D., and Guinness, F. E. (1986). Great expectations: dominance, breeding success and offspring sex ratios in red deer. *Animal Behaviour* 34, 460–471.

Clutton-Brock, T. H., Albon, S. D., and Guinness, F. E. (1988). Reproductive success in male and female red deer. In T. H. Clutton-Brock, ed., *Reproductive Success: Studies of Individual Variation in Contrasting Breeding Systems.* Chicago, IL: The University of Chicago Press, pp. 325–343.

Clutton-Brock, T. H., Guinness, F. E., and Albon, S. D. (1982). *Red Deer: Behavior and Ecology of Two Sexes.* Chicago, IL: The University of Chicago Press.

Clutton-Brock, T. H., Harvey, P. H., and Rudder, B. (1977). Sexual dimorphism, socioeconomic sex ratio and body weight in primates, *Nature* 269, 797–799.

Codde, S. A., Allen, S. G., Houser, D. S., and Crocker, D. E. (2016). Effects of environmental variables on surface temperature of breeding female northern elephant seals, *Mirounga angustirostris*, and pups. *Journal of Thermal Biology* 61, 98–105.

Condit, R. and Le Boeuf, B. J. (1984). Feeding habits and feeding grounds of the northern elephant seal. *Journal of Mammalogy* 65, 281–290.

Condit, R., Reiter, J., Morris, P. A., Berger, R., Allen, S. G. and Le Boeuf, B. J. (2014). Lifetime survival and senescence of northern elephant seals. *Marine Mammal Science* 30, 122–138.

Cooke, F. and Rockwell, R. F. (1988). Reproductive success in a lesser snow goose population. In T. H. Clutton-Brock, ed., *Reproductive Success: Studies of Individual Variation in Contrasting Breeding Systems.* Chicago, IL: University of Chicago Press, pp. 237–250.

Costa, D. P., Breed, G. A., and Robinson, P. W. (2012). New insights into pelagic migrations: implications for ecology and conservation. *Annual Review of Ecology, Evolution, and Systematics* 43, 73–96.

Costa, D. P., Crocker, D. E., Croll, D., Goley, D. A., Houser, D. P., Le Boeuf, B. J., Waples, D., Webb, P., and Calambokidis, J. (1996). Effects of the California ATOC experiment on marine mammals. *Journal of the Acoustic Society of America* 100, 2581.

Costa, D. P., Crocker, D. E., Gedamke, J., Webb, P. M., Houser, D. S., Blackwell, S. B., Waples, D., Hayes, S. A., and Le Boeuf, B. J. (2003). The effect of a low-frequency sound source (acoustic thermometry of the ocean climate) on the diving behavior of juvenile northern elephant seals, *Mirounga angustirostris. Journal of the Acoustic Society of America* 11, 1155–1165.

Costa, D. P., Crocker, D. E., Waples, D. M., Webb, P. M., Gedamke, J., Houser, D. S., Goley, D. P., Le Boeuf, B. J., and Calambokidis, J. (1998). The California marine mammal research program of the acoustic thermometry of ocean climate experiment: potential effects of low frequency sound on the distribution and behavior of marine mammals. *California and the World Ocean Symposium volume* 2, pp. 1542–1553.

Costa, D. P., Le Boeuf, B. J., Huntley, A. C., and Ortiz, C. L. (1986). The energetics of lactation in the northern elephant seal, *Mirounga angustirostris*. *Journal of Zoology* 209, 21–33.

Côté, S. D. and Festa-Bianchet, M. (2001). Reproductive success in female mountain goats: the influence of age and social rank. *Animal Behaviour* 62, 173–181.

Cox, C. R. and Le Boeuf. B. J. (1977). Female incitation of male competition: a mechanism in sexual selection. *The American Naturalist* 111, 317–335.

Crocker, D. E., Costa, D. P., Le Boeuf, B. J., Webb, P. M., and Houser, D. S. (2006). Impact of El Niño on the foraging behavior of female northern elephant seals. *Marine Ecology Progress Series* 309, 1–10.

Crocker, D. E., Houser, D. S., and Webb, P. M. (2012). Impact of body reserves on energy expenditure, water flux, and mating success in breeding male northern elephant seals. *Physiological and Biochemical Zoology* 85, 11–20.

Crocker, D. E., Le Boeuf, B. J., and Costa, D. P. (1997). Drift diving in female northern elephant seals: implications for food processing. *Canadian Journal of Zoology* 75, 27–39.

Crocker, D. E., Le Boeuf, B. J., Naito, Y., Asaga, T., and Costa, D. P. (1994). Swim speed and dive function in a female northern elephant seal. In B. J. Le Boeuf and R. Laws, eds., *Elephant Seals: Population Ecology, Behavior, and Physiology*. Berkeley: University of California Press, pp. 328–342.

Crocker, D. E., Williams, J. D., Costa, D. P., and Le Boeuf, B. J. (2001). Maternal traits and reproductive effort in northern elephant seals. *Ecology* 82, 3541–3555.

Croll, D. A., Acevedo-Gutiérrez, A., Tershy, B. R., and Urbán-Ramírez, J. (2001). The diving behavior of blue and fin whales: is dive duration shorter than expected based on oxygen stores? *Comparative Biochemistry and Physiology Part A* 129, 797–809.

Daneri, G. and Carlini, A. (2002). Fish prey of southern elephant seals, *Mirounga leonina*, at King George Island. *Polar Biology* 25, 739–743.

Daneri, G., Carlini, A., and Rodhouse, P. (2000). Cephalopod diet of the southern elephant seal, *Mirounga leonina*, at King George Island, South Shetland Islands. *Antarctic Science* 12, 16–19.

David, J. H. M. (1987). South African fur seal, *Arctocephalus pusillus pusillus*. In J. P. Croxall and R. L. Gentry, eds., *Status, Biology and Ecology of Fur Seals*. NOAA Technical Report NMFS No. 51. Princeton, NJ: Princeton University Press, pp. 65–72.

Davies, N. B. (1991). Mating systems. In J. R. Krebs and N. B. Davies, eds., *Behavioural Ecology* (3rd ed.). London: Blackwell Scientific, pp. 263–299.

Davis, R. W. and Weihs, D. (2007). Locomotion in diving elephant seals: physical and physiological constraints. *Philosophical Transactions on Royal Society B* 362, 2141–2150.

Davis, R. W., Fuiman, L. A., Williams, T. M., and Le Boeuf, B. J. (2001). Three-dimensional movements and swimming activity of a northern elephant seal. *Comparative Biochemistry and Physiology - Part A: Molecular & Integrative Physiology* 129, 759–770.

Debier, C., Chalon, C., Le Boeuf, B. J., de Tillesse, T., Larondelle, Y., and Thomé, J. (2006). Mobilization of PCBs from blubber to blood in northern elephant seals (*Mirounga angustirostris*) during the post-weaning fast. *Aquatic Toxicology* 80, 149–157.

Debier, C., Le Boeuf, B. J., Ikonomou, M. G., de Tillesse, T., Larondelle, Y., and Ross, P. S. (2005a). Polychlorinated biphenyls, dioxins, and furans in weaned, free-ranging northern elephant seal pups from Central California, USA. *Environmental Toxicology and Chemistry* 24, 629–633.

Debier, C., Ylitalo, G. M., Weise, M., Gulland, F., Costa, D. P., Le Boeuf, B. J., de Tillesse, T., and Larondelle, Y. (2005b). PCBs and DDT in the serum of juvenile California sea lions: associations with vitamins A and E and thyroid hormones. *Environmental Pollution* 134, 323–332.

DeLong, R. L. and Stewart, B. S. (1991). Diving patterns of northern elephant seal bulls. *Marine Mammal Science* 7, 369–384.

DeLong, R. L., Gilmartin, W. G., and Simpson, J. G. (1973). Premature births in California sea lions: association with high organochlorine pollutant residue levels. *Science* 181, 1168–1170.

Desprez, M., Harcourt, R., Hindell, M. A., Cubaynes, S., Gimenez, O., and McMahon, C. R. (2014). Age-specific cost of first reproduction in female southern elephant seals. *Biological Letters* 10, 20140264.

Deutsch, C. J., Crocker, D. E., Costa, D. P., and Le Boeuf, B. J. (1994). Sex and age related variation in reproductive effort of northern elephant seals. In B. J. Le Boeuf and R. Laws, eds., *Elephant Seals: Population Ecology, Behavior, and Physiology*. Berkeley: University of California Press, pp. 169–210.

Deutsch, C. J., Haley, M. P., and Le Boeuf, B. J. (1990). Reproductive effort of male northern elephant seals: estimates from mass loss. *Canadian Journal of Zoology* 68, 2580–2593.

Dobson, F. S. and Michener, G. R. (1995). Maternal traits and reproduction in Richardson's ground squirrels. *Ecology* 76, 851–862.

Downhower, J. F. and Armitage, K. B. (1971). The yellow-bellied marmot and the evolution of polygamy. *American Nature* 105, 355–370.

Ducatez, S., Dalloyau, S., Richard, P., Guinet, C., and Cherel, Y. (2008). Stable isotopes document winter trophic ecology and maternal investment of adult female southern elephant seals (*Mirounga leonina*) breeding at the Kerguelen Islands. *Marine Biology* 155, 413–420.

Eder, E. B., Lewis, M. N., Campagna, C., and Koch, P. L. (2010). Evidence of demersal foraging from stable isotope analysis of juvenile elephant seals from Patagonia. *Marine Mammal Science* 26, 430–442.

Elorriaga-Verplancken, F. R., Blanco-Jarvio, A., Silva-Segundo, C. A., Paniagua-Mendoza, A., Rosales-Nanduca, H., Robles-Hernández, R., Mote-Herrera, S., José Amador-Capitanachi, M., and Sandoval-Sierra, J. (2020). A southern elephant seal (*Mirounga leonina*) in the Gulf of California: genetic confirmation of the northernmost record to date. *Aquatic Mammals* 46, 137–145.

Elsner, R. and Gooden, B. (1983). *Diving and Asphyxia*. Cambridge: Cambridge University Press.

Emlen, S. T. and Oring, L. W. (1977). Ecology, sexual selection, and the evolution of mating systems. *Science* 197, 215–223.

Ericsson, G., Wallin, K., Ball, J. P., and Broberg, M. (2001). Age-related reproductive effort and senescence in free-ranging moose, *Alces alces*. *Ecology* 82, 1613–1620.

Falke, K., Hill, R. D., Qvist, J., Schneider, R. C., Guppy, M., Liggins, G. C., Hochachka, P. W., Elliott, R. E., and Zapol, W. M. (1985). Seal lungs collapse during free diving: evidence from arterial nitrogen tensions. *Science* 229, 556–558.

Fay, F. H., Ray, G. C., and Kibal'chich, A. A. (1984). Time and location of mating and associated behaviour of the Pacific walrus, *Odobenus rosmarus divergens* Illiger. In Soviet–American Cooperative Research on Marine Mammals: Vol. 1. Pinnipeds: 89–99. NOAA Technical Report NMFS 12, US Department of Commerce, National Marine Fisheries Service.

Fedak, M. A. and Anderson, S. S. (1982). The energetics of lactation: accurate measurements from a large wild mammal, the grey seal (*Halichoerus grypus*). *Journal of Zoology* 198, 473–479.

Fedak, M. A., Arnbom, T., and Boyd, I. L. (1997). Factors affecting maternal expenditure in southern elephant seals during lactation. *Ecology* 79, 471–483.

Fedak, M. A., Pullen, M. R., and Kanwisher, J. (1988). Circulatory responses of seals to periodic breathing: heart rate and breathing during exercise and diving in the laboratory and open sea. *Canadian Journal of Zoology* 66, 53–60.

Field, I. C., Bradshaw, C. J. A., Burton, H. R., Sumner, M. D., and Hindell, M. A. (2005). Resource partitioning through oceanic segregation of foraging juvenile southern elephant seals (*Mirounga leonina*). *Oecologia* 142, 127–135.

Field, I., Bradshaw, C., van den Hoff, J., Burton, H., and Hindell, M. (2007). Age-related shifts in the diet composition of southern elephant seals expand overall foraging niche. *Marine Biology* 150, 1441–1452.

Field, I., Hindell, M., Slip, D., and Michael, K. (2001). Foraging strategies of southern elephant seals (*Mirounga leonina*) in relation to frontal zones and water masses. *Antarctic Science* 13, 371–379.

Fletcher, S., Le Boeuf, B. J., Costa, D. P., Tyack, P. L., and Blackwell, S. B. (1996). Onboard acoustic recording from diving northern elephant seals. *Journal of the Acoustical Society of America* 100, 2531–2539.

Fowler, G. S. (1995). Stages of age-related reproductive success in birds: simultaneous effects of age, pair-bond duration and reproductive experience. *American Zoology* 35, 318–328.

Franklin, W. L., Mossman, A. S., and Dole, M. (1975). Social organization and home range of Roosevelt elk. *Journal of Mammal* 56, 102–118.

Gadgil, M. and Bossert, W. H. (1970). Life historical consequences of natural selection. *American Nature* 104, 1–24.

Gaillard, J. M., Allainé, D., Pontier, D., Yoccz, N. G. I., and Promislow, E. L. (1994). Senescence in natural populations of mammals: a reanalysis. *Evolution* 489, 509–516.

Gaillard, J. M., Festa-Bianchet, M., Delorme, D., and Jorgenson, J. (2000). Body mass and individual fitness in female ungulates: bigger is not always better. *Proceedings of the Royal Society of London B* 267, 471–477

Garcia-Aguilar, M. C., Turrent, C., Elorriaga-Verplancken, F. R., Arias-Del-Razo, A., and Schramm, Y. (2018). Climate change and the northern elephant seal (*Mirounga angustirostris*) population in Baja California, Mexico. *PLoS ONE* 13(2), e0193211.

Gaskin, D. E. (1982). *The Ecology of Whales and Dolphins*. London: Heinemann.

Gentry, R. (1998). *Behavior and Ecology of the Northern Fur Seal*. Princeton, NJ: Princeton University Press.

Gentry, R. L. and Kooyman, G. L. (1986). Eds., *Fur Seals: Maternal Strategies on Land and at Sea*. Princeton, NJ: Princeton University Press.

Geoffroy Saint-Hilaire, É. and Cuvier, F. (1826). Phoques (vol. 39, p. 540–559) *In* Dictionnaire de sciénces naturelles … Strasbourg and Paris, 60 vols. + atlas of 12 vols., 1816–1830.

Gill, T. (1866). On a new species of the genus *Macrorhinus*. *Proceedings of the Chicago Academic of Sciences* 1, 33–34.

Glassow, M. A. (1980). Recent developments in the archaeology of the Channel Islands. In D. M. Power, ed., *The California Islands: Proceedings of a Multidisciplinary Symposium*. Santa Barbara, CA: Santa Barbara Museum of Natural History, pp. 79–99.

Goetsch, C., Conners, M. G., Budge, S. M., Mitani, Y., Walker, W. A., Bromaghin, J. F., Simmons, S. E., Reichmuth, C., and Costa, D. P. (2018). Energy-rich mesopelagic fishes revealed as a critical prey resource for a deep-diving predator using quantitative fatty acid signature analysis. *Frontiers in Marine Science* 5, 1–19.

Gray, J. E. (1827). Mammalia (vol. 5, 391 p). *In The animal kingdom arranged in conformity with its organization, by the Baron (G.) Cuvier, with additional descriptions ... by Edward Griffith ... and others*. London, George B. Whittaker 16 vols., 1827–1835.

Gray, D. (1882). Notes on the characteristics and habits of the bottlenose whale (*Hyperoodon rostratus*). *Proceedings of the Zoological Society of London* 50, 726–731.

Green, K. and Burton, H. (1993). Comparison of the stomach contents of southern elephant seals, *Mirounga leonina*, at Macquarie and Heard Islands. *Marine Mammal Science* 9, 10–22.

Guhl, A. M. (1962). The behaviour of chickens. In F. S. E. Hafez, ed., *The Behavior of Domestic Animals*. London: Bailliere, Tindal and Cox, pp. 491–530.

Hacker, E. S. (1986). Stomach content analysis of short-finned pilot whales (*Globicephala macrorhynchus*) and northern elephant seals (*Mirounga angustirostris*) from the southern California Bight. Southwest Fisheries Center, La Jolla, CA. Administrative Rep. NO. LJ-86-08C.

Hakoyama, H., Le Boeuf, B. J., Naito, Y., and Sakamoto, W. (1994). Diving behavior in relation to ambient water temperature in northern elephant seals. *Canadian Journal of Zoology* 72, 643–651.

Haley, M. P., Deutsch, C. J., and Le Boeuf, B. J. (1991). A method for estimating mass of large pinnipeds. *Marine Mammal Science* 7, 157–164.

Hall, K. R. L. (1968). Behaviour and ecology of the wild patas monkey, *Erythrocebus patas*, in Uganda. In P. Jay, ed., *Primates: Studies in Adaptation and Variability*. New York: Holt, Rinehart and Winston, pp. 120–130.

Hamilton, W. D. (1964). The genetical theory of social behavior. I, II. *Journal of Theoretical Biology* 7, 1–52.

Hammond, P. S., McConnell, B. J., and Fedak, M. A. (1993). Grey seals off the east coast of Britain: distribution and movements at sea. In I. L. Boyd, ed., Marine Mammals: advances in Behavioural and Population Biology. *Symposia of the Zoological Society of London* 66, 211–224.

Hanni, K. D. and Pyle, P. (2000). Entanglement of pinnipeds in synthetic materials at south-east Farallon Island, California, 1976-1998. *Marine Pollution Bulletin* 40, 1076–1081.

Hansen, S. and Lavigne, D. M. (1997). Temperature effects on the breeding distribution of grey seals (*Halichoerus grypus*). *Physiological Zoology* 70(4), 436–443.

Hassrick, J. L., Crocker, D. E., Zeno, R. L., Blackwell, S. B., Costa, D. P., and Le Boeuf, B. J. (2007). Swimming speed and foraging strategies of northern elephant seals. *Deep Sea Research Part II: Topical Studies in Oceanography* 54, 369–383.

Hassrick, J. L., Crocker, D. E., Teutschel, N. M., McDonald, B. I., Robinson, P. W., Simmons, S. E., and Costa, D. P. (2010). Condition and mass impact oxygen stores and dive duration in adult female northern elephant seals. *Journal of Experimental Biology* 213, 585–592.

Hawkes, K. and Smith, K. R. (2010). Do women stop early? Similarities in fertility decline in humans and chimpanzees. *Annual New York Academy Science* 1204, 43–53.

Hawkes, K., Smith, K. R., and Blevins, J. K. (2012). Human actuarial ageing increases faster when background death rates are lower: a consequence of differential heterogeneity? *Evolution* 66, 103–114.

Heath, C. B. and Francis, J. M. (1983). Breeding behavior in the California sea lion. Mugu Lagoon and San Nicolas Island. *Ecological Research Symposium* 3, 145–150.

Heath, C. B. and Francis, J. M. (1987). Mechanisms and consequences of mate choice in the California sea lion. *Biennial Conference on the Biology of Marine Mammalogy* (Abstract).

Heath, M. E. and Schusterman, R. J. (1975). "Displacement" sand flipping in the northern elephant seal (*Mirounga angustirostris*). *Behavioral Biology* 1, 379–385.

Hill, S. E. B. (1987). Reproductive ecology of Weddell seals (*Leptonychotes weddelli*) in McMurdo Sound, Antarctica. PhD thesis, University of Minnesota, St. Paul, MN.

Heezen, B. C. (1957). Whales entangled in deep sea cables. *Deep-Sea Research* 4, 105–115.

Heide-Jørgensen, M., Laidre, K. L., Nielsen, N. H., Hansen, R. G., and Røstad, A. (2013). Winter and spring diving behavior of bowhead whales relative to prey. *Animal Biotelemetry* 1, 15

Hill, S. E. B. (1987). Reproductive ecology of Weddell seals (*Leptonychotes weddelli*) in McMurdo Sound, Antarctica. Ph.D. thesis, University of Minnesota, St. Paul, MN.

Hindell, M. A. (1991). Some life-history parameters of a declining population of southern elephant seals. Mirounga leonina. *Journal of Animal Ecology* 60, 119–134.

Hindell, M. A. and Burton, H. R. (1987). Past and present status of the southern elephant seal (*Mirounga leonina*) at Macquarie Island. *Journal of Zoology* 213, 365–380.

Hindell, M. A. and Lea, M. A. (1998). Heart rate, swimming speed and estimated oxygen consumption of a free-ranging southern elephant seal. *Physiological Zoology* 71, 74–84.

Hindell, M. A., Slip, D. J., and Burton, H. R. (1991). The diving behavior of adult male and female southern elephant seals, *Mirounga leonina* (Pinnipedia: Phocidae). *Australian Journal of Zoology* 39, 595–619.

Hoelzel, A. R. (1999). Impact of population bottlenecks on genetic variation and the importance of life-history; a case study of the northern elephant seal. In P. A. Racey, P. J. Bacon, J. F. Dallas, and S. B. Piertney, eds., Molecular Genetics in Animal Ecology. *Biological Journal of the Linnean Society* 68, 23–39.

Hoelzel, A. R. and Le Boeuf, B. J. (1990). Low variation in northern elephant seal fingerprints. *Fingerprint News* 2, 10.

Hoelzel, A. R., Fleisher, R. C., Campagna, C., Le Boeuf, B. J., and Alvord, G. (2002). Impact of a population bottleneck on symmetry and genetic diversity in the northern elephant seal. *Journal of Evolutionary Biology* 15, 567–575.

Hoelzel, A. R., Halley, J., O'Brien, S. J., Campagna, C., Arnborm, T., Le Boeuf, B. J., Ralls, K., and Dover, G. A. (1993). Elephant seal genetic variation and the use of simulation models to investigate historical population bottlenecks. *Journal of Heredity* 84, 443–449.

Holekamp, K. E., Smale, L., and Szykman, M. (1996). Rank and reproduction in the female spotted hyaena. *Journal of Reproduction and Fertility* 108, 229–237.

Hooker, S. K. and Baird, R. W. (1999). Deep-diving behaviour of the northern bottlenose whale, *Hyperoodon ampullatus* (Cetacea: Ziphiidae). *Proceedings of the Royal Society of London B* 266, 671–676.

Hooker, S. K. and Baird, R. W. (2001). Diving and ranging behaviour of odontocetes: a methodological review and critique. *Mammalian Review* 31, 81–105.

Holland, K. N., Brill, R. W., Chang, R. K. C., Sibert, J. R., and Fournier, D. A. (1992). Physiological and behavioral thermoregulation in bigeye tuna (*Thunnus obesus*). *Nature* 358, 410–412.

Hoogland, J. L. (1995). *The Black-Tailed Prairie Dog*. Chicago, IL: The University of Chicago Press.

Hoogland, J. L. and Sherman, P. W. (1976). Advantages and disadvantages of Bank Swallow (*Riparia riparia*) coloniality. *Ecology Monograph* 46, 33–58.

Hoyt, E. (1992). Whale watching around the world. *International Whale Bulletin* 7, 1–8.

Hoyt, E. (1993). Saving whales by watching them. *New Scientist* 138, 45–46.

Hoyt, E. (2003). Toward a new ethic for watching dolphins and whales. In T. Frohoff and B. Peterson, eds., *Between Species: Celebrating the Dolphin/Human Bond*. San Francisco, CA: Sierra Club Books, pp. 168–177.

Hrdy, S. B. (1999). *Mother Nature: A History of Mothers, Infants, and Natural Selection*. New York: Pantheon Books.

Huber, H. R., Beckham, C., and Nisbet, J. (1991). Effects of the 1982-83 el Niño on northern elephant seals on the South Farallon Islands, California. In F. Trillmich and K. A. Ono, eds., *Pinnipeds and El Niño: Responses to Environmental Stress*. Berlin: Springer Verlag, pp. 219–233.

Huber, S., Millesi, E., Walzl, M., Dittami, J., and Arnold, W. (1999). Reproductive effort and costs of reproduction in female European ground squirrels. *Oecologia* 121, 19–24.

Hückstädt, L. A., Koch, P. L, McDonald, B. I., Goebel, M. E., Crocker, D. E., and Costa, D. P. (2012). Stable isotope analyses reveal individual variability in the trophic ecology of a top marine predator, the southern elephant seal. *Oecologia* 169, 395–406.

Hückstädt, L. A. et al. (2018). The extra burden of motherhood: reduced dive duration associated with pregnancy status in a deep-diving mammal, the northern elephant seal. *Biology Letters* 14(2), 20170722.

Huntley, A. C. and Costa, D. P. (1983). Cessation of ventilation during sleep: a unique mode of energy conservation in the northern elephant seal. *Proceedings of the International Union Physiology Science* 15, 203.

Huntley, A. C., Costa, D. P., and Rubin, R. D. (1984). The contribution of nasal countercurrent heat exchange to water balance in the northern elephant seal, *Mirounga angustirostris*. *Journal of Experimental Biology* 113, 447–454.

Jarman, P. J. (1974). The social organization of antelope in relation to their ecology. *Behaviour* 4, 215–267.

Jepsen, E. M. and Nico de Bruyn, P. J. (2019). Pinniped entanglement in oceanic plastic pollution: a global review. *Marine Pollution Bulletin* 145, 295–305.

Johnson, D. (1990). A southern elephant seal (*Mirounga leonina*) in the northern hemisphere (Sultanate of Oman). *Marine Mammal Science* 12, 242–243.

Jones, O. R., Scheuerlein, A., Salguero-Gomez, R., Camarda, C. G., Schaible, R., Casper, B. B. et al. (2014). Diversity of ageing across the tree of life. *Nature* 505, 169–173.

Jonker, F. C. and Bester, M. N. (2004). Seasonal movements and foraging areas of adult southern female elephant seals, *Mirounga leonina*, from Marion Island. *Antarctic Science* 10, 21–30.

Judson, O. (2002). *Dr. Tatiana's Sex Advice to All Creation*. New York: Henry Holt and Company.

Kannan, K., Kajiwara, N., Le Boeuf, B. J., and Tanabe, S. (2004). Organochlorine pesticides and polychlorinated biphenyls in California sea lions. *Environmental Pollution* 3, 425–434.

Kastak, D. and Schusterman, R. J. (1998). Low-frequency amphibious hearing in pinnipeds: methods, measurements, noise, and ecology. *Journal of the Acoustical Society of America* 103, 2216–2228.

Kendal-Barr, J. M., Vyssotski, A. L., Mukhametov, L. M., Siegel, J. M., and Lyamin, O. I. (2019). Eye state asymmetry during aquatic unihemispheric slow wave sleep in northern fur seals (*Callorhinus ursinus*). *PLoS ONE* 14, e0217025.

Kenney, R. (1979). Breathing and heart rates of the southern elephant seal, *Mirounga leonina* L. *Papers and Proceedings of the Royal Society of Tasmanian* 113, 21–27.

Kienle, S. S. (2019). Intraspecific variation and behavioral flexibility in the foraging strategies of seals. PhD thesis, University of California at Santa Cruz.

Kiyota, M., Baba, N., and Mouri, M. (1992). Occurrence of northern elephant seal in Japan. *Marine Mammal Science* 8, 433.

Kleiber, M. (1932). Body size and metabolism. *Hilgardia* 6, 315–351.

Kleiman, D. G. (1977). Monogamy in mammals. *Quarterly Review of Biology* 52, 39–69.

Klimley, A. P., Le Boeuf, B. J., Cantara, K. M., Richert, J. E., Davis, S. F., Van Sommeran, S., and Kelly, J. T. (2001). The hunting strategy of white sharks (*Carcharodon carcharias*) near a seal colony. *Marine Biology* 138, 617–636.

Kooyman, G. L. (1965). Techniques used in measuring diving capacities of Weddell seals. *Polar Records* 12, 391–394.

Kooyman, G. L. (1981). *Weddell Seal: Consummate Diver.* Cambridge: Cambridge University Press.

Kooyman, G. L. (1985). Physiology without restraint in diving mammals. *Marine Mammal Science* 1, 166–178.

Kooyman, G. L. (1989). *Diverse Divers: Physiology and Behaviors.* Berlin: Springer Verlag.

Kooyman, G. L., Kerem, D. H., Campbell, W. B., and Wright, J. J. (1973). Pulmonary gas exchange in freely diving Weddell seals, *Leptonychotes weddelli*. *Respiratory Physiology* 17, 283–290.

Kovacs, K. M. (1989). Mating strategies of male hooded seals. 8th Biennial Conference on the Biology of Marine Mammals. Pacific Grove, California Dec 7–11, 1989 (abstract).

Kramer, D. L. (1988). The behavioral ecology of air breathing by aquatic animals. *Canadian Journal of Zoology* 66, 89–94.

Kretzmann, M. B., Costa, D. P., and Le Boeuf, B. J. (1993). Maternal energy investment in elephant seal pups: evidence for sexual equality? *The American Naturalist* 141, 466–480.

Laidre, K. L., Heide-Jørgensen, M., Dietz, R., Hobbs, R. C., and Jørgensen, O. A. (2003). Deep-diving by narwhals *Monodon monoceros*: differences in foraging behavior between wintering areas? *Marine Ecology Progress Series* 261, 269–281.

Lawick-Goodall, J. V. (1967). *My Friends the Wild Chimpanzees.* Washington: National Geographic Society.

Lawick-Goodall, J. V. (1971). *In the Shadow of Man.* Boston: Houghton Mifflin Company.

Laws, R. M. (1953). The elephant seal (*Mirounga leonina* Linn.), part 1: growth and age. *Falkland Islands Dependencies Survey, London* 8, 1–62.

Laws, R. M. (1956). The elephant seal (*Mirounga leonina* Linn.), II. General, social, and reproductive behaviour. Falkland Islands Dependencies Survey, *London Scientific Reports* 13, 1–88.

Le Boeuf, B. J. (1971a). The aggression of the breeding bulls. *Natural History* 130, 82–94.

Le Boeuf, B. J. (1971b). Oil contamination and elephant seal mortality: a "negative" finding. In D. Straughan, ed., *Biological and Oceanographical Survey of the Santa Barbara Channel Oil Spill, 1969–1970*, volume I. Los Angeles, CA: Allan Hancock Foundation, University of Southern California, pp. 277–285.

Le Boeuf, B. J. (1972). Sexual behavior in the northern elephant seal, *Mirounga angustirostris*. *Behaviour* 41, 1–26.

Le Boeuf, B. J. (1974). Male–male competition and reproductive success in elephant seals. *American Zoologist* 14, 163–176.

Le Boeuf, B. J. (1977). Back from extation? *Pacific Discovery* 30, 1–9.

Le Boeuf, B. J. (1981). History. In B. J. Le Boeuf and S. Kaza, eds., *The Natural History of Año Nuevo*. Pacific Grove, CA: The Boxwood Press, pp. 1–60.

Le Boeuf, B. J. (1984). Beach warfare: sexual conflict among northern elephant seals. In D. Macdonald, ed., *The Encyclopedia of Mammals*. New York: Facts on File, Inc., pp. 284–285.

Le Boeuf, B. J. (1991). Pinniped mating systems on land, ice, and in the water: emphasis on the phocidae. In D. Renouf, ed., *Behavior of Pinnipeds*. New York: Chapman and Hall, pp. 45–65.

Le Boeuf, B. J. (1994). Variation in the diving pattern of northern elephant seals with age, mass, sex, and reproductive condition. In B. J. Le Boeuf and R. Laws, eds., *Elephant Seals: Population Ecology, Behavior and Physiology*. Berkeley: University of California Press, pp. 237–252.

Le Boeuf, B. J. (1995). Behavioral issues in returning marine mammals to their habitat. In D. J. St. Aubin, J. R. Geraci, and Lounsbury, V. J., eds., *Rescue, Rehabilitation and Release of Marine Mammals: An Analysis of Current Views and Practices*. Contract # T75136433, Marine Mammal Commission, Proceedings of a Workshop held Dec. 3–5, 1991, Des Plaines, IL, pp. 62–70.

Le Boeuf, B. J. (2002). Status of pinnipeds on Santa Catalina Island. *Proceedings of the California Academy of Sciences* 53, 11–21.

Le Boeuf, B. J. (2004). Hunting and migratory movements of white sharks in the eastern north Pacific. *Memoirs of National Institute of Polar Research* Special Issue, 58, 91–102.

Le Boeuf, B. J. and Bonnell, M. L. (1971). DDT in California sea lions. *Nature* 234, 108–110.

Le Boeuf, B. J. and Briggs, K. T. (1977). The cost of living in a seal harem. *Mammalia* 41, 167–196.

Le Boeuf, B. J. and Campagna, C. (1993). Protection and abuse of young in pinnipeds. In S. Parmigiani and F. vom Saal, eds., *Infanticide and Parental Care*. London: Harwood Academic Publishers, pp. 257–276.

Le Boeuf, B. J. and Campagna, C. (2013). Wildlife viewing spectacles: best practices from elephant seal (*Mirounga* sp.) colonies. *Aquatic Mammals* 39(2), 132–146.

Le Boeuf, B. J. and Condit, R. (1983). The high cost of living on the beach. *Pacific Discovery* 36, 12–14.

Le Boeuf, B. J. and Crocker, D. E. (1996). Diving behavior of elephant seals: implications for predator avoidance. In P. Klimley and D. Ainley, eds., *Great White Sharks: The Biology of Carcharadon carcharias*. New York: Academic Press, pp. 193–205

Le Boeuf, B. J. and Crocker, D. E. (2005). Ocean climate and seal condition. *BMC Biology* 3, 9.

Le Boeuf, B. J. and Kaza, S. (eds.) (1981). *The Natural History of Año Nuevo*. Pacific Grove, CA: Boxwood Press.

Le Boeuf, B. J. and Laws, R. M. (1994). Elephant seals: an introduction to the genus. In B. J. Le Boeuf and R. M. Laws, eds., *Elephant Seals: Population Ecology, Behavior and Physiology*, Berkeley, CA: University of California Press, pp. 1–16.

Le Boeuf, B. J. and Mesnick, S. (1991). Sexual behavior of male northern elephant seals: I. lethal injuries to adult females. *Behaviour* 116, 143–162.

Le Boeuf, B. J. and Naito, Y. (in press). Dive types matter: they reveal the foraging ecology of elephant seals. In D. P. Costa and E. McHuron, eds., *Ethology of Earless Seals*. Berlin: Springer-Verlag.

Le Boeuf, B. J., and Ortiz, C. L. (1977). Composition of elephant seal milk. *Journal of Mammalogy* 58, 683–685.

Le Boeuf, B. J., and Panken, K. J. (1977). Elephant seals breeding on the mainland in California. *Proceedings of the California Academy of Sciences*, 41, 267–280.

Le Boeuf, B. J. and Peterson, R. S. (1969a). Dialects in elephant seals. *Science* 166, 1654–1656.

Le Boeuf, B. J. and Peterson, R. S. (1969b). Social status and mating activity in elephant seals. *Science* 163, 91–93.

Le Boeuf, B. J. and Petrinovich, L. F. (1974a). Elephant seals: interspecific comparisons of vocal and reproductive behavior. *Mammalia* 38, 16–32.

Le Boeuf, B. J. and Petrinovich, L. F. (1974b). Dialects of northern elephant seals, *Mirounga angustirostris*: origin and reliability. *Animal Behaviour* 22, 656–663.

Le Boeuf, B. J. and Petrinovich, L. F. (1975). Elephant seal dialects: are they reliable? *Rapports et procès-verbaux des réunions, Conseil Permanent International pour L'exploration de la Mer* 169, 213–218.

Le Boeuf, B. J. and Reiter, J. (1988). Lifetime reproductive success in northern elephant seals. In T. H. Clutton-Brock, ed. *Reproductive Success: Studies of Individual Variation in Contrasting Breeding Systems*. Chicago, IL: University of Chicago Press, pp. 344–362

Le Boeuf, B. J. and Reiter, J. (1991). Biological effects associated with El Niño, Southern Oscillation 1982–83, on northern elephant seals breeding at Año Nuevo, California. In F. Trillmich and K. A. Ono, eds., *Pinnipeds and El Niño: Responses to Environmental Stress*. Berlin: Springer Verlag, pp. 206–218.

Le Boeuf, B. J., Ainley, D. G., and Lewis, T. J. (1974). Elephant seals on the Farallones: population structure of an incipient breeding colony. *Journal of Mammalogy* 55, 370–385.

Le Boeuf, B. J., Condit, R., and Reiter, J. (1989). Parental investment and the secondary sex ratio in northern elephant seals. *Behavioral Ecology and Sociobiology* 25, 109–117.

Le Boeuf, B. J., Condit, R., and Reiter, J. (2019). Lifetime reproductive success of northern elephant seals (*Mirounga angustirostris*). *Canadian Journal of Zoology* 97, 1203–1217.

Le Boeuf, B. J., McCosker, J., and Hewitt, J. (1987). Crater wounds on northern elephant seals: the cookie cutter shark strikes again. *Fishery Bulletin* 85, 387–392

Le Boeuf, B. J., Morris, P., and Reiter, J. (1994). Juvenile survivorship of northern elephant seals from Año Nuevo. In B. J. Le Boeuf and R. M. Laws, eds., *Elephant Seals: Population Ecology, Behavior, and Physiology*.Berkeley: University of California Press, pp. 121–136.

Le Boeuf, B. J., Riedman, M., and Keyes, R. S. (1982b). White shark predation on pinnipeds in California coastal waters. *Fishery Bulletin* 80, 891–895.

Le Boeuf, B. J., Whiting, R. J., and Gantt, R. F. (1972). Perinatal behavior of northern elephant seal females and their young. *Behaviour* 43, 121–156.

Le Boeuf, B. J., Costa, D. P., Huntley, A. C., Kooyman, G. L., and Davis, R. W. (1986). Pattern and depth of dives in northern elephant seals, *Mirounga angustirostris*. *Journal of Zoology* 208, 1–7.

Le Boeuf, B. J., Costa, D. P., Huntley, A. C., and Feldcamp. S. D. (1988). Continuous, deep diving in female northern elephant seals, *Mirounga angustirostris*. *Canadian Journal of Zoology* 66, 446–458.

Le Boeuf, B. J., Naito, Y., Huntley, A. C., and Asaga, T. (1989). Prolonged, continuous, deep diving by northern elephant seals. *Canadian Journal of Zoology* 67, 2514–2519.

Le Boeuf, B. J., Naito, Y., Asaga, T., Crocker, D. E., and Costa, D. P. (1992). Swim speed in female northern elephant seal: metabolic and foraging implications. *Canadian Journal of Zoology* 70, 786–795.

Le Boeuf, B. J., Crocker, D. E., Blackwell, S. B., Morris, P. A., and Thorson, P. H. (1993). Sex differences in diving and foraging behavior of northern elephant seals. *Symposia of the Zoological Society of London* 66, 149–178.

Le Boeuf, B. J., Morris, P. A., Blackwell, S. B., Crocker, D. E., and Costa, D. P. (1996). Diving behavior of juvenile northern elephant seals. *Canadian Journal of Zoology* 74, 1632–1644.

Le Boeuf, B. J., Crocker, D. E., Costa, D. P., Blackwell, S. B., Webb, P. M., and Houser, D. S. (2000a). Foraging ecology of northern elephant seals. *Ecological Monographs* 70, 353–382.

Le Boeuf, B. J., Crocker, D. E., Grayson, J., Gedamke, J., Webb, P. M., Blackwell, S. B., and Costa, D. P. (2000b). Respiration and heart rate at the surface between dives in northern elephant seals. *Journal of Experimental Biology* 203, 3265–3274.

Le Boeuf, B. J., Condit, R., Morris, P. A., and Reiter, J. (2011). The northern elephant seal rookery at Año Nuevo: a case study in colonization. *Aquatic Mammals* 37, 486–501.

Le Boeuf, B. J., Giesy, J. P. Kannan, K., Kajiwara, N., Tanabe, S., and Debier, C. (2002). Organochloride pesticides in California sea lions revisited. *BMC Ecology* 2, 11.

Lee, P. C., Fishlock, V., Webber, C. E., and Moss, C. J. (2016). The reproductive advantages of a long life: longevity and senescence in wild female African elephants. *Behavioral Ecology and Sociobiology* 70, 337–345.

Lemaitre, J. F., Ronget, V., Tidière, M., Allainé, D., Berger, V., Cohas, A., Colchero, F., Conde, D. A., Garratt, M., Liker, A. Marais, G. A. B., Scheuerlein, A., Székely, T., and Gaillard, J. M. (2020). Sex differences in adult lifespan and aging rates of mortality across wild animals. *PNAS* 117, 8546–8553.

Levenson, D. H. and Schusterman, R. J. (1997). Pupillometry in seals and sea lions: ecological implications. *Canadian Journal of Zoology* 75, 2050–2057.

Levenson, D. H. and Schusterman, R. J. (1999). Dark adaptation and visual sensitivity in shallow and deep-diving pinnipeds. *Marine Mammal Science* 15, 1303–1313.

Lewis, R., O'Connell, T. C., Lewis, M., Campagna, C., and Hoelzel, A. R. (2006). Sex-specific foraging strategies and resource partitioning in the southern elephant seal (*Mirounga leonina*). *Proceedings of the Royal Society B* 273, 2901–2907.

Lieberg-Clark, P., Bacon, C. E., Burns, S. A., Jarman, W. M., and Le Boeuf. B. J. (1995). DDT in California sea-lions: a follow-up study after twenty years. *Marine Pollution Bulletin* 30, 744–755.

Linnaeus, C. (1758). *Systema naturæ per regna tria naturæ, secundum classes, ordines, genera, species, cum characteribus, differentiis, synonymis*, locis. **1** (10th ed.). Stockholm: Laurentius Salvius. pp. [1–4], 1–824.

Lockyer, C. (1977). Observations on diving behaviour of the sperm whale, *Physeter catodon*. In M. Angel, ed., *A Voyage of Discovery*. Oxford: Pergamon, pp. 591–609.

Loison, A., Festa-Bianchet, M., Gaillard, J-M., Jorgenson, J. T., and Jullien, J-M. (1999). Age-specific survival in five populations of ungulates: evidence of senescence. *Ecology* 80, 2539–2554.

Lowry, M. S., Condit, R., Hatfield, B., Allen, S. G., Berger, R., Morris, P. A., Le Boeuf, B. J., and Reiter, J. (2014). Abundance, distribution, and population growth of the northern elephant seal (*Mirounga angustirostris*) in the United States from 1991 to 2010. *Aquatic Mammals* 40(1), 36–47.

Lowry, M. S., Jaime, E. M., Nehasil, S. E., Betcher, A., and Condit, R. (2020). Winter surveys at the Channel Islands and Point Conception reveal population growth of northern elephant seals and residence counts of other pinnipeds, U.S. Department of Commerce, NOAA Technical Memorandum NMFS-SWFSC-627.

Luenser, K., Fickel, J., Lehnen, A., Speck, S., and Ludwig, A. (2005). Low level of genetic variability in European bisons (*Bison bonasus*) from the Bia, Mowieza National Park in Poland. *European Journal of Wildlife Research* 51(2), 84–87.

Lyamin, O. I., Mukhametov, L. M., Siegel, J. M., Nazarenko, E. A., Polyakova, I. G., and Shpak, O. V. (2002). Unihemispheric slow wave sleep and the state of the eyes in a white whale. *Behavioral Brain Research* 129, 125–129.

Lyamin, O. I., Pryaslova, J., Kosenko, P. O., and Siegel, J. M. (2007). Behavioural aspects of sleep in bottlenose dolphin mothers and their calves. *Physiology & Behavior* 92, 725–733.

Lyamin, O. I., Kosenko, P. O., Vyssotski, A. L., Lapierre, J. L., Siegel, J. M., and Mukhametov, L. M. (2012). Study of sleep in a walrus. *Doklady Biological Sciences* 444, 188–191.

Lyamin, O. I., Kosenko, P. O., Korneva, S. M., Vyssotski, A., Mukhametov, M., and Siegel, J. M. (2018). Fur seals suppress REM sleep for very long periods without subsequent rebound. *Current Biology* 28, 2000–2005.

Maaswinkel, H. and Whishaw, I. Q. (1999). Homing with locale, taxon, and dead reckoning strategies by foraging rats: sensory hierarchy in spatial navigation. *Behavioral Brain Research* 99, 143–152.

Matsumura, M., Watanabe, Y. Y., Robinson, P. W., Miller, P. J. O., Costa, D. P., and Miyazaki, N. (2011). Underwater and surface behavior of homing juvenile northern elephant seals. *Journal of Experimental Biology* 214, 629–636.

MacDonald, D. W. (1983). The ecology of carnivore social behaviour. *Nature* 301, 379–384.

McCann, T. S. (1981). Aggression and sexual activity of male southern elephant seals, Mirounga leonina. *Journal of Zoology* 195, 295–310.

McCann, T. S. (1982). Aggressive and maternal activities of female southern elephant seals (*Mirounga leonina*). *Animal Behaviour* 30, 268–276.

McCann, T. S., Fedak, M. A., and Harwood, J. (1989). Parental investment in southern elephant seals, Mirounga leonina. *Behavioural Ecology and Sociobiology* 25, 81–87.

McCleery, R. H., Perrins, C. M., Sheldon, B. C., and Charmantier, A. (2008). Age-specific reproduction in a long-lived species: the combined effects of senescence and individual quality. *Proceedings of the Royal Society B* 275, 963–970.

McGovern, K. A., Rodriguez, D. H., Lewis, M. N., and Davis, R. W. (2019). Classification and behavior of free-ranging female southern elephant seal dives based on three-dimensional movements and video-recorded observations. *Marine Ecology Progress Series* 620, 215–232.

McIntyre, T., de Bruyn, P. J. N., Ansorge, I. J., Bester, M. N., Bornemann, H., Plötz, J., and Tosh, C. A. (2010). A lifetime at depth: vertical distribution of southern elephant seals in the water column. *Polar Biology* 33, 1037–1048.

McIntyre, T., Bornemann, H., Plötz, J., Tosh, C. A., and Bester, M. N. (2011). Water column use and forage strategies of female southern elephant seals from Marion Island. *Marine Biology* 158, 2125–2139.

McIntyre, T., Bornemann, H., Plötz, J., Tosh, C. A., and Bester, M. N. et al. (2012). Deep divers in even deeper seas: habitat use of male southern elephant seals from Marion Island. *Antarctic Science* 24, 561–570.

McLean, I. G. and Towns, A. J. (1981). Differences in weight changes and the annual cycle of male and female arctic ground squirrels. *Arctic* 34, 249–254.

Melin, S. R., Orr, A. J., and DeLong, R. L. (2008). The status of the northern fur seal population at San Miguel Island, California, 2006 and 2007. In J. W. Testa, ed., *Fur Seal Investigations, 2006–2007*. U.S. Dep. Commer., NOAA Tech. Memo. NMFS-AFSC-188, pp. 41–54.

Mesnick, S. L. and Le Boeuf, B. J. (1991). Sexual behavior of male northern elephant seals: II. female response to potentially injurious encounters. *Behaviour* 117, 262–280.

Mesnick, S. L., del Carmen Garcia Rivas, M., Le Boeuf, B. J., and Peterson, S. M. (1998). Northern elephant seals in the gulf of California, México. *Marine Mammal Science* 14, 171–178.

Michener, G. R. (1984). Sexual differences in body weight patterns of Richardson's ground squirrels during the breeding season. *Journal of Mammalogy* 65, 59–66.

Miersch, L., Hanke, W., Wieskotten, S., Hanke, F. D., Oeffner, J., Leder., A. Brede, M., Witte, M., and Dehnhardt, G. (2011). Flow sensing by pinniped whiskers. *Philosophical Transactions of the Royal Society B* 366, 3077–3084.

Milinkovitch, M. C., Kanitz, R., Tiedermann, R., Tapia W., Llerena, F., Caccone, A., Gibbs, K., and Powell, J. (2013). Recovery of a nearly extinct Galapagos tortoise despite minimal genetic variation. *Evolutionary Applications* 6(2), 377–383.

Miller, P. J. O., Aoki, K., Rendell, L. E., and Amano, M. (2008). Stereotypical resting behavior of the sperm whale. *Current Biology* 18, R21–R23.

Minamikawa, S., Iwasaki, T., and Kishiro, T. (2007). Diving behaviour of a Baird's beaked whale, *Berardius bairdii*, in the slope water region of the western North Pacific: first dive records using a data logger. *Fisheries Oceanography* 16, 573–577.

Miller, K. W. (1972). Inert gas narcosis and animals under high pressure. *Symposia of the Society of Experimental Biology* 26, 363–378.

Mitani, Y., Andrews, R. D., Sato, K., Kato, A., Naito, Y., and Costa, D. P. (2010). Three-dimensional resting behaviour of northern elephant seals: drifting like a falling leaf. *Biology Letters* 6(2), 163–166.

Morton, M. L. and Sherman, P. W. (1978). Effects of a spring snowstorm on behavior, reproduction and survival of Belding's ground squirrels. *Canadian Journal of Zoology* 56, 2578–2590.

Mukhametov, L. M., Supin, A. Y., and Polyakova, I. G. (1977). Interhemispheric asymmetry of the electroencephalographic sleep patterns in dolphins. *Brain Research* 134, 581–584.

Munk, W., Worcester, P., and Wunsch, C. (1995). *Ocean Acoustic Tomography*. Cambridge: Cambridge University Press.

Müller, J. and Henle, F. G. J. (1839). *Systematische Beschreibung der Plagiostomen*. Berlin: Veit & Co.

Naito, Y., Costa, D. P., Adachi, T., Robinson, P. W., Fowler, M., and Takahashi, A. (2013). Unravelling the mysteries of a mesopelagic diet: a large apex predator specializes on small prey. *Functional Ecology* 27(3), 710–717.

Naito, Y., Costa, D. P., Adachi, T., Robinson, P. W., Peterson, S. H., Mitani, Y., and Takahashi, A. (2017). Oxygen minimum zone: an important oceanographic habitat for deep-diving northern elephant seals, *Mirounga angustirostris*. *Ecology and Evolution* 7(16), 6259–6270.

Naito, Y., Le Boeuf, B. J., Asaga, T., and Huntley, A. C. (1989). Long-term diving records of an adult female northern elephant seal. *Nankyoku Shiryo* (Polar Rec.) 33, 1–9.

Neuhaus, P. (2000). Weight comparisons and litter size manipulation in Columbian ground squirrels (*Spermophilus columbianus*) show evidence of costs of reproduction. *Behavioral Ecology and Sociobiology*, 48, 75–83.

Newland, C., Field, I. C., Nichols, P. D., Bradshaw, C. J. A., and Hindell, M. A. (2009). Blubber fatty acid profiles indicate dietary resource partitioning between adult and juvenile southern elephant seals. *Marine Ecology Progress Series* 384, 303–312.

Nottebohm, F. (1969). The song of the chingolo, *Zonotrichia capensis* in Argentina: description and evaluation of a system of dialects. *Condor* 71, 299–315.

Nottebohm, F. (1972). Ontogeny of bird song. *Science* 167, 950–956.

Nussey, D. H., Froy, H., Lemaitre, J. F., Gaillard, J. M., and Austad, S. N. (2013). Senescence in natural populations of animals: widespread evidence and it implications for bio-gerontology. *Ageing Research Review* 12, 214–225.

Nussey, D. H., Kruuk, L. E. B., Donald, A., Fowlie, M., and Clutton-Brock, T. H. (2006). The rate of senescence in maternal performance increases with early-life fecundity in red deer. *Ecology Letters* 9, 1342–1350.

O'Brien, S., Roelke, M., Marker, L., Newman, A., Winkler, C., Meltzer, D., Colly, L., Evermann, J., Bush, M., and Wildt, D. (1985). Genetic basis for species vulnerability in the Cheetah. *Science* 227, 1428–1434.

Oftedal, O. T., Boness, D. J., and Bowen, W. D. (1988). The composition of hooded seal (*Cystophora cristata*) milk: an adaptation for postnatal fattening. *Canadian Journal of Zoology* 66(2), 318–322.

Okello, M. M., Manka, S. G., and D'Amour, D. E. (2008). The relative importance of large mammal species for tourism in Amboseli National Park, Kenya. *Tourism Management* 29, 751–760.

Oliver, G. W., Morris, P. A., Thorson, P. H., and Le Boeuf, B. J. (1998). Homing behavior of juvenile northern elephant seals. *Marine Mammal Science* 14, 245–256.

Ollason, J. C. and Dunnet, G. M. (1988). Variation ion breeding success in fulmars. In T. H. Clutton-Brock, ed., *Reproductive Success: Studies of Individual Variation in Contrasting Breeding Systems*. Chicago, IL: University of Chicago Press, pp. 263–278

Olsson, M. and Shine, R. (1996). Does reproductive success increase with age or with size in species with indeterminate growth? A case study using sand lizards (*Lacerta agilis*). *Oecologia* 105, 175–178.

Orgeret, F., Cox, S. L., Weimerskirch, H., and Guinet, C. (2018). Body condition influences ontogeny of foraging behavior in juvenile southern elephant seals. *Ecology and Evolution* 9, 223–236.

Orians, G. H. (1969). On the evolution of mating systems in birds and mammals. *American Naturalist* 103, 257–264.

Oring, L. W. and Knudson, M. L. (1973). Monogamy and polyandry in the spotted sandpiper. *The Living Bird* 11, 5973.

Oring, L. W. and Lank, D. B. (1982). Sexual selection, arrival times, philopatry and site fidelity in the polyandrous spotted sandpiper. *Behavioral Ecology and Sociobiology* 10, 185–191.

Orr, R. T. and Poulter, T. C. (1965). The pinniped population of Año Nuevo Island, California. *Proceedings of the California Academy of Sciences* 32, 377–404.

Orr, R. T. and Poulter, T. C. (1967). Some observation on reproduction, growth, and social behavior in the Steller sea lion. *Proceedings of the California Academy of Sciences* 35, 193–226.

Ortiz, C. L., Costa, D. P., and Le Boeuf, B. J. (1978). Water and energy flux in elephant seal pups fasting under natural conditions. *Physiological Zoology* 51, 166–178.

Ortiz, C. L., Le Boeuf, B. J., and Costa, D. P. (1984). Milk intake of elephant seal pups: an index of parental investment. *The American Naturalist* 124, 416–422.

O'Shea, T. J. and Brownell, R. (1998). California sea lion (Zalophus californianus) populations and Σ DDT contamination. *Marine Pollution Bulletin* 36, 159–164.

Packer, C., Herbst, L., Pusey, A. E., Bygott, J. D., Hanby, J. P., Cairns, S. J., and Borgerhoff Mulder, M. (1988). Reproductive success in lions. In T. H. Clutton-Brock, ed., *Reproductive Success: Studies of Individual Variation in Contrasting Breeding Systems*. Chicago, IL: University of Chicago Press, pp. 363–383.

Packer, C., Tatar, M., and Collins, A. (1998). Reproductive cessation in female mammals. *Nature* 392, 807–811.

Papi, F. (1992). *Animal Homing*. London: Chapman and Hall.

Payne, R. B. (1984). Sexual Selection, Lek and Arena Behavior and Sexual Size Dimorphism in Birds. *Ornithological Monographs* no. 33. Washington, DC., American Ornithologists' Union.

Peterson, R. S. (1965). Behavior of the northern fur seal. PhD thesis, Johns Hopkins university, Baltimore, MD.

Peterson, R. S., Le Boeuf, B. J., and DeLong, R. L. (1968). Fur seals from the Bering Sea breeding in California. *Nature* 219, 899–901.

Pianka, E. R. and Parker, W. S. (1975). Age-specific reproductive tactics. *American Nature* 109, 453–464.

Piatkowski, U., Vergani, D., and Stanganelli, Z. (2002). Changes in the cephalopod diet of southern elephant seal females at King George Island, during El Niño–La Niña events. *Journal of the Marine Biological Association of the United Kingdom* 82, 913–916.

Pistorius, P. A., Bester, M. N., and Kirkman, S. P. (1999). Survivorship of a declining population of southern elephant seals. *Mirounga leonina*, in relation to age, sex and cohort. *Oecologia* 121, 210–211.

Pistorius, P. A., Bester, M. N., Lewis, M. N., Taylor, F. E., Campagna, C., and Kirkman, S. P. (2004). Adult female survival, population trend, and the implications of early primiparity in a capital breeder, the southern elephant seal (*Mirounga leonina*). *Journal of Zoology*, 263, 107–119.

Price, T. (1998). Maternal and paternal effects in birds, In T. A. Mousseau and C. W. Fox, eds., *Maternal Effects as Adaptations*. New York: Oxford University Press, pp. 202–226.

Promislow, D. E. L. (1991). Senescence in natural populations of mammals: a comparative study. *Evolution* 45, 1869–1887.

Quick, N. J., Cioffi, W. R., Shearer, J. M., Fahlman, A., and Read, A. J. (2020). Extreme diving in mammals: first estimates of behavioral aerobic dive limits in Cuvier's beaked whales. *Journal of Experimental Biology* 225, 222109.

Quinn, T. P. and Branton, E. L. (1982). The use of celestial and magnetic cues by orienting sockeye salmon smolts. *Journal of Comparative Physiology* 147, 547–552.

Radford, K. W., Orr, R. T., and Hubbs, C. L. (1965). Reestablishment of the northern elephant seal (*Mirounga angustirostris*) off central California. *Proceedings of the California Academy of Sciences* 31, 601–612.

Ralls, K. (1976). Mammals in which females are larger than males. *The Quarterly Review of Biology* 51, 245–276.

Rea, L. D. and Costa, D. P. (1992). Changes in standard metabolism during long-term fasting in northern elephant seal pups (*Mirounga angustirostris*). *Physiological Zoology* 65, 97–111.

Rebke, M., Coulson, T., Becker, P. H., and Vaupel, J. W. (2010). Reproductive improvement and senescence in a long-lived bird. *PNAS* 107, 7841–7846.

Redwood, S. and Felix, F. (2018). The most northerly record of a southern elephant seal (*Mirounga leonina*) in the Pacific Ocean at the island of Taboga, Gulf of Panama, Panama. *Aquatic Mammals* 44(1), 13–18.

Reid, J. M., Bignal, E. M., Bignal, S., McCracken, D. I., and Monaghan, P. (2003). Age-specific reproductive performance in red-billed choughs Pyrrhocorax pyrrhocorax: patterns and processes in a natural population. *Journal of Animal Ecology* 72, 765–776.

Rose, M. R. (1991). *Evolutionary Biology of Aging*. New York: Oxford University Press.

Reiter, J. and Le Boeuf, B. J. (1991). Life history consequences of variation in age at primiparity in northern elephant seals. *Behavioral Ecology and Sociobiology* 28, 153–160.

Reiter, J., Panken, K. J., and Le Boeuf, B. J. (1981). Female competition and reproductive success in northern elephant seals. *Animal Behaviour* 29, 670–687.

Reiter, J., Stinson, N. L., and Le Boeuf, B. J. (1978). Northern elephant seal development: the transition from weaning to nutritional independence. *Behavioral Ecology and Sociobiology* 3, 337–367.

Ridgeway, S. (2002). Asymmetry and symmetry in brain waves from dolphin left and right hemispheres: some observations after anesthesia during quiescent hanging behavior, and during visual obstruction. *Brain Behavior Evolution* 60, 265–274.

Riedman, M. L. (1982). The evolution of alloparental care and adoption in mammals and birds. *The Quarterly Review of Biology* 57(4), 405–435.

Riedman, M. L. (1990). *The Pinnipeds: Seals, Sea Lions, and Walruses*. Berkeley: University of California Press.

Riedman, M. L. and Le Boeuf, B. J. (1982). Mother–pup separation and adoption in northern elephant seals. *Behavioral Ecology and Sociobiology* 11, 203–215.

Riedman, M. L. and Ortiz, C. L. (1979). Changes in milk composition during lactation in the northern elephant seal. *Physiology Zoology* 52, 240–249.

Rieger, J. F. (1996). Body size, litter size, timing of reproduction, and juvenile survival in the Uinta ground squirrel, *Spermophilus armatus*. *Oecologia* 107, 463–468.

Riofrio-Lazo, M., Aurioles-Gamboa, D., and Le Boeuf, B. J. (2012). Ontogenetic changes in feeding habits of northern elephant seals revealed by $\delta15N$ and $\delta13C$ analysis of growth layers in teeth. *Marine Ecology Progress Series* 450, 229–241.

Rivarola, M., Campagna, C., and Tagliorette, A. (2001). Demand-driven commercial whale watching in Península Valdés (Patagonia): conservation implications for right whales. *Journal of Cetacean Research Management (Special Issue)* 2, 145–151.

Robbins, C. T., Podbielancik-Norman, R. S., Wilson, D. L., and Mould, E. D. (1981). Growth and nutrient consumption of elk calves compared to other ungulate species. *Journal of Wildlife Management* 45, 172–186.

Robinson, P. W., Simmons, S. E., Crocker, D. E., and Costa, D. P. (2010). Measurements of foraging success in a highly pelagic marine predator, the northern elephant seal. *Journal of Animal Ecology* 79(6), 1146–1156.

Robinson, P. W., Costa, D. P., Crocker, D. E., Gallo-Reynoso, J. P., Champagne, C. D., and Fowler, M. A. (2012). Foraging behavior and success of a mesopelagic predator in the northeast Pacific Ocean: insights from a data-rich species, the northern elephant seal. *PLoS ONE* 7(5), e36728.

Rose, N. A., Deutsch, C. J., and Le Boeuf, B. J. (1991). Sexual behavior of male northern elephant seals: III. The mounting of weaned pups. *Behaviour* 119, 171–192.

Rowley, I. and Russell, E. (1990). Splendid fairy-wrens: demonstrating the importance of longevity. In P. B. Stacey and W. D. Koenig, eds., *Cooperative Breeding in Birds: Long-*

Term Studies of Ecology and Behavior. Cambridge: Cambridge University Press, pp. 1–30.

Rule, J. P., Adams, J. W., Marx, F. G., Evans, A. R., Tennyson, A. N. D., Scofield, R. P., and Fitzgerald, E. M. G. (2020). First monk seal from the Southern Hemisphere rewrites the evolutionary history of true seals. *Proceedings of the Royal Society B* 28720202318.

Rutberg, A. T. (1983). The evolution of monogamy in primates. *Journal of Theoretical Biology* 104, 93–112.

Saether, B. E. (1990). Age-specific variation in reproductive performance of birds. *Current Ornithology* 7, 251–283.

Saijo, D., Mitani, Y., Abe, T., Sasaki, H., Goetsch, C., Costa, D. P., and Miyashita, K. (2017). Linking mesopelagic prey abundance and distribution to the foraging behavior of a deep-diving predator, the northern elephant seal. *Deep-Sea Research Part Ii – Topical Studies in Oceanography* 140, 163–170.

Salogni, E., Sanvito, S., and Galimberti, F. (2015). Postmortem examination and causes of death of northern elephant seal (*Mirounga angustirostris*) pups at the San Benito Islands, Baja California, Mexico. *Marine Mammal Science* 32(2), 743–752.

Samuelson, J. (1890). *India, Past and Present, Historical, Social, and Political.* London: Trübner & Co.

Scammon, C. M. (1874). *The Marine Mammals of the Northwestern Coast of North America.* San Francisco, CA: J.H. Carmany and Co.

Schmidt-Nielsen, K. (1972). Locomotion: energy cost of swimming, flying and running. *Science* 177, 222–228.

Schmidt-Nielsen, K. (1981). Countercurrent systems in animals. *Scientific American* 244, 118–128.

Schorr, G. S., Falcone, E. A., Moretti, D. J., and Andrews, R. D. (2014). First long-term behavioral records from Cuvier's beaked whales (*Ziphius cavirostris*) reveal record-breaking dives. *PLoS ONE* 9, e92633.

Secor, S. M. (2009). Specific dynamic action: a review of the postprandial metabolic response. *Journal of Comparative Physiology B* 179, 1–56.

Secor, S. M. and Diamond, J. (1998). A vertebrate model of extreme physiological regulation *Nature* 395, 659–662

Selighsohn, E. M. (1987). Dominance relationships and reproductive success within bands of feral ponies. Ph.D. thesis, University of Connecticut, Storrs.

Sharp, S. P. and Clutton-Brock, T. H. (2010). Reproductive senescence in a cooperatively breeding mammal. *Journal of Animal Ecology* 79, 176–183.

Shearer, J. M., Quick, N. J., Cioffi, W. R., Baird, R. W., Webster, D. L., Foley, H. J., Swaim, Z. T., Waples, D. M., Bell, J. T., and Read, A. J. (2019). Diving behaviour of Cuvier's beaked whales (*Ziphius cavirostris*) off Cape Hatteras, North Carolina. *Royal Society of Open Science* 6, 181728.

Simon, M., Johnson, M., Tyack, P., and Madsen, P. T. (2009). Behaviour and kinematics of continuous ram filtration in bowhead whales (*Balaena mysticetus*). *Proceedings of the Royal Society B* 276, 1–10.

Simmons, S. E., Crocker, D. E., Kudela, R. M., and Costa, D. P. (2007). Linking foraging behaviour of the northern elephant seal with oceanography and bathymetry at mesoscales. *Marine Ecology Progress Series* 346, 265–275.

Skibiel, A. L., Dobson, F. S., and Murie, J. O. (2009). Maternal influences on reproduction in two populations of Columbian ground squirrels. *Ecology Monograph* 79, 325–341.

Slip, D. J. (1997). Foraging ecology of southern elephant seals from Heard Island. PhD thesis, University of Tasmania, Tasmania.

Southall, B. L., Casey, C., Holt, M., Insley, S., and Reichmuth, C. (2019). High-amplitude vocalizations of male northern elephant seals and associated ambient noise on a breeding colony. *Journal of the Acoustical Society of America* 146, 4514–4524.

Stearns, S. C. (1976). Life-history tactics: a review of the ideas. *Quarterly Review of Biology* 51, 3–47.

Stephens, P. A., Houston, A. I., Harding, K. C., Boyd, I. L., and McNamara, J. M. (2014). Capital and income breeding: the role of food supply. *Ecology* 95(4), 882–896.

Stewart, B. S. and Yochem, P. K. (1991). Northern elephant seals on the Southern California Channel Islands and El Niño. In F. Trillmich and K. A. Ono, eds., *Pinnipeds and El Niño: Responses to Environmental Stress*. Berlin: Springer Verlag, pp. 234–243.

Stewart, B. S., Yochem, P. K., Le Boeuf, B. J., Huber, H. R., DeLong, R. L., Jameson, R. J., Sydeman, W. J., and Allen, S. G. (1994). Population recovery and status of the northern elephant seal *Mirounga angustirostris*. In B. J. Le Boeuf and R. M. Laws, eds., *Elephant Seals: Population Ecology, Behavior, and Physiology*. Berkeley: University of California Press, pp. 29–48.

Stirling, I. (1975). Factors affecting the evolution of social behaviour in the pinnipeds. *Rapports et procès-verbaux des réunions Conseil Permanent International pour L'exploration de la Mer* 169, 205–212.

Skibiel, A. L., Dobson, F. S., and Murie, J. O. (2009). Maternal influences on reproduction in two populations of Columbian ground squirrels. *Ecology Monograph* 79, 325–341.

Sydeman, W. J. and Nur, N. (1994). Life history strategies of female northern elephant seals. In B. J. Le Boeuf and R. M. Laws, eds., *Elephant Seals: Population Ecology, Behavior, and Physiology*. Berkeley: University of California Press, pp. 121–136.

Sydeman, W. J., Huber, H., Emslie, S. D., Ribic, C. A., and Nur, N. (1991). Age-specific weaning success of northern elephant seals in relation to previous breeding experience. *Ecology* 72, 2204–2217.

Tapper, R. (2006). *Wildlife Watching and Tourism: A Study of the Benefits and Risks of a Fast Growing Tourism Activity and Its Impacts on Species*. Bonn, Germany: United Nations Environment Programme (UNEP)/Convention on Migratory Species K (CMS).

Thielcke, G. (1969). Geographic variation in bird vocalizations. In R. A. Hinde, ed., *Bird Vocalizations: Their Relations to Current Problems in Biology and Psychology*. London: Cambridge University Press.

Thomas, F., Teriokhin, A. T., Renaud, T. F. DeMeeûs, T., and Guégan, J. F. (2000). Human longevity at the cost of reproductive success: evidence from global data. *Journal of Evolution Biology* 13, 409–414.

Thompson, M. E., Jones, J. H., Pusey, A. E., Brewer-Marsden, S., Goodall, J., Marsden, D., Matsuzawa, T., and Wrangham, R. W. (2007). Aging and fertility patterns in wild chimpanzees provide insights into the evolution of menopause. *Current Biology* 17, 2150–2156.

Thornton, S. J., Pelc, N. J., Spielman, D. M., Liao, J. R., Costa, D. P., Crocker, D. E. et al. (1997a). Vascular flow dynamics in a diving elephant seal (*Mirounga angustirostris*). *International Society for Magnetic Resonance in Medicine* 2, 823.

Thornton, S. J., Spielman, D. M., Block, W. F., Hochachka, P. W., Crocker, D. E., Le Boeuf, B. J. et al. (1997b). Imaging in a diving seal. *International Society for Magnetic Resonance in Medicine* 2, 822.

Thornton, S. J., Spielman, D. M., Pelc, N. J., Block, W. F., Crocker, D. E., Costa, D. P., Le Boeuf, B. J., and Hochachka, P. W. (2001). Effects of forced diving on the spleen and hepatic sinus in northern elephant seal pups. *Proceedings of the National Academy of Sciences of the United States of America* 98, 9413–9418.

Thornton, S. J., Hochachka, P. W., Crocker, D. E., Costa, D. P., Le Boeuf, B. J., Spielman, D. M., and Pelc, N. J. (2005). Stroke volume and cardiac output in juvenile elephant seals during forced dives. *Journal of Experimental Biology* 208, 3637–3643.

Thorson, P. H. and Le Boeuf, B. J. (1994). Developmental aspects of diving in northern elephant seal pups. In B. J. Le Boeuf and R. M. Laws, eds., *Elephant Seals: Population Ecology, Behavior, and Physiology*. Berkeley: University of California Press, pp. 271–289

Townsend, C. H. (1912). The northern elephant seal *Macrorhinus angustirostris* (Gill). *Zoologica* 1, 159–173.

Tricas, T. C. and McCosker, J. E. (1984). Predatory behavior of the white shark (*Carcharodon carcharias*, with notes on its biology. *Proceedings of the California Academy of Sciences* 43 (14), 221–238.

Trivers, R. L. (1972). Parental investment and sexual selection, In B. Campbell, ed., *Sexual Selection and the Descent of Man: 1871–1971*. London: Heinemann, pp. 135–179.

Trivers, R. L. and Willard, D. E. (1973). Natural selection of parental ability to vary the sex ratio of offspring. *Science* 179, 90–92.

Valenzuela-Toro, A. and Pyenson, N. D. (2019). What do we know about the fossil record of pinnipeds? A historiographical investigation. *Royal Society of Open Science* 6, 191394.

Valenzuela-Toro, A. M., Gutstein, C. A., Varas-Malca, R. M., Suarez, M. E., and Pyenson, N. D. (2013). Pinniped turnover in the south Pacific Ocean: new evidence from the PliO-Pleistocene of the Atacama Desert, Chile. *Journal of Vertebrate Paleontology* 33, 216–223.

Valenzuela-Toro, A. M., Zicos, M. H., and Pyenson, N. D. (2020). Extreme dispersal or human-transport? The enigmatic case of an extralimital freshwater occurrence of a southern elephant seal from Indiana. *Peer J* 8, e9665.

van de Pol, M. and Verhulst, S. (2006). Age-dependent traits: a new statistical model to separate within-and between-individuals effects. *American Nature* 167, 766–773.

Vyssotski, A. L., Dell'Omo, G., Dell'Ariccia, G, Abramchuk, A. N., Serkov, A. N., Latanov, A. V., Loizzo, A., Wolfer, D. P., and Lipp, H. P. (2009). EEG responses to visual landmarks in flying pigeons. *Current Biology* 19, 1159–1166.

Walters, A., Lea, M. A., van den Hoff, J., Field, I. C., Virtue, P., Sokolov, S., Pinkerton, M. H., and Hindell, M. A. (2014). Spatially explicit estimates of prey consumption reveal a new krill predator in the Southern Ocean. *PLoS ONE* 9(1), e86452.

Watkins, W. A., Daher, M. A., Fristrup, K. M. Howald, T. J., and Di Sciara, G. N. (1993). Tagged with transponders and tracked underwater by Sonar. *Marine Mammal Science* 1, 55–67.

Weihs, D. (1973). Mechanically efficient swimming techniques for fish with negative buoyancy. *Journal of Marine Research* 31, 194–209.

Weladji, R. B., Gaillard, J. M., Yoccoz, N. G., Holand, Ø., Mysterud, A., Loison, A., Nieminen, M., and Stenseth, N. C. (2006). Good reindeer mothers live longer and become better in raising offspring. *Proceedings of the Royal Society of London B* 273, 1239–1244.

Weladji, R. B., Holand, Ø., Gaillard, J-M., Yoccoz, N. G., Mysterud, A., Nieminen, M., and Stenseth, N. C. (2010). Age-specific changes in different components of reproductive output in female reindeer: terminal allocation or senescence? *Oecologia* 162, 261–271.

White, F. N. and Odell, D. K. (1971). Thermoregulatory behavior of the northern elephant seal, *Mirounga angustirostris. Journal of Mammals* 52, 758–774.

Williams, G. C. (1966). *Adaptation and Natural Selection*. Princeton, NJ: Princeton University Press

Williams, T. M., Davis, R. W., Fuiman, L. A., Francis, J., Le Boeuf, B. J., Horning, M., Calambokidis, J., and Croll, D. A. (2000). Sink or swim: strategies for cost-efficient diving by marine mammals. *Science* 288, 133–136.

Wilson, E. O. (1975). *Sociobiology*. Cambridge, MA: Harvard, Belknap Press.

Wilson, E. O. and Bossert, W. H. (1971). *A Primer of Population Biology*. Stamford, CT: Sinauer Associates, Inc.

Worthy, G. A. J., Morris, P. A., Costa, D. P., and Le Boeuf, B. J. (1992). Moult energetics of the *northern elephant seal (Mirounga angustirostris). Journal of Zoology* 227, 257–265.

Xirocostas, Z. A., Everingham, S. E., and Moles, A. T. (2010). The sex with the reduced sex chromosome dies earlier: a comparison across the tree of life. *Biological Letters* 16, 20190867.

Yorio, P., Frere, E., Gandini, P., and Schiavini, A. (2001). Tourism and recreation at seabird breeding sites in Patagonia, Argentina: current concerns and future prospects. *Bird Conservation International* 11, 231–245.

Yoshino, K., Takahashi, A., Adachi, T., Costa, D. P., Robinson, P. W., Peterson, S. H., Hückstädt, L. A., Holser, R. R., and Naito, Y. (2020). Acceleration-triggered animal-borne videos show a dominance of fish in the diet of female northern elephant seals. *Journal of Experimental Biology* 223, jeb212936.

Zhang, Ya-Ping, Wang, Xiao-xia, Ryder, Oliver A., Li, Hai-Peng, Zhang, He-Ming, Yong, Yange, and Wang, Peng-yan. (2002). Genetic diversity and conservation of endangered animal species. *Pure and Applied Chemistry* 74(4), 575–584.

Index

Printed in the United States
by Baker & Taylor Publisher Services